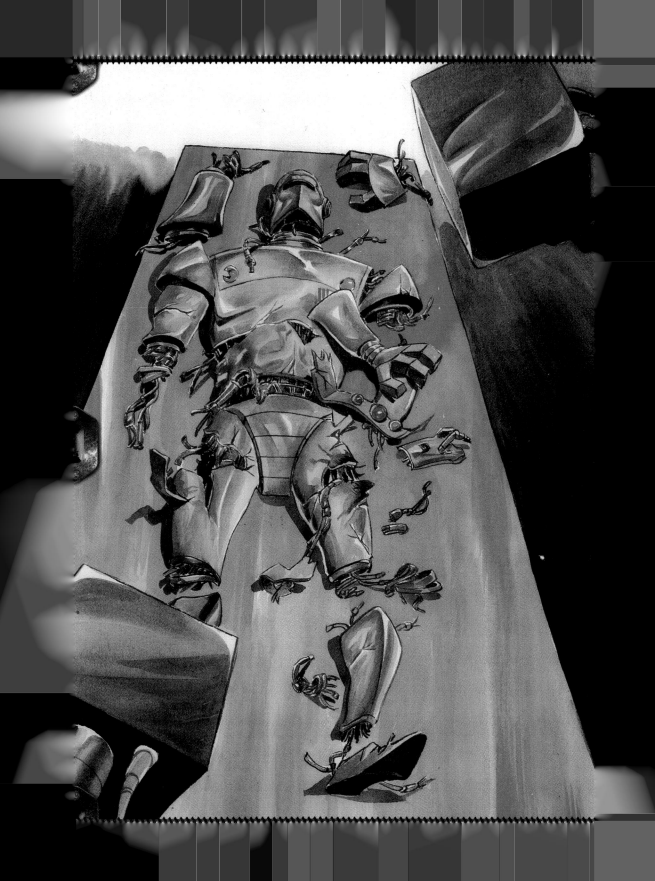

"It's on your head, then. Let's go."

"Wait."

"You just do your work, Sonny Jim. Nothing funny. I'll be right here. Three hours." Fenniman sealed the door.

As tumblers fell, Zac whispered to the robot, too softly for Fenniman or microphones to hear: "Robillard."

A spark leaped across Golden Boy's eyes, and a soft whir caught Fenniman's attention.

"What did you do?"

"Shh." Then, to the robot: "All systems off. All systems!"

"Break it?" Fenniman said.

"It was low on power, anyway. A few decades in the ground will do that to you. But if it's got a bomb or something in it, it shouldn't be a problem now."

"You're sure of that?"

"Well…I can hope. Got a can opener?" He selected a screwdriver and soldering pen from the wall. Methodically, he heated the seals on the robot's chest and temples, and pried at the seams. Time crawled, until Zac's shoulders and wrists hurt, and Fenniman yawned uncontrollably, boredom cracking his emotionless veneer.

After more than two hours, the plates fell away, opening the robot up. Zac wiped the perspiration from his eyes and stared. It was a glimpse not only of the past, but of his ancestor's unsuspected genius. The shell was there for looks and added protection, but nothing in the body structure suggested a more intrinsic need for it. The "muscle" system was as simple as a child's constructor set, rods and hydraulic shafts in an elaborate pulley system, united by electrical cords and small bulbs at the joints that Zac recognized as elementary solenoids.

"They should have patented this," he muttered. "They'd both have been rich."

"Find something?" Fenniman stared at him dumbly.

"A bit of history, that's all. Robotics is far beyond this now, but this technology is stunning for the 40s."

The golden skull peeled away. A small pool of liquid leaked onto Zac's fingers, giving him a faint shock. After the body's secrets, he had expected to see something resembling a microcomputer in the head, another precursor of what he understood to be robotics, but there was only a block of moist gray foam.

He squeezed it between thumb and forefinger, and the damp gray gave gently. Again a faint shock tingled his fingers. The mass was electric. Wires poked into it, a one-way circuit.

Even through the concrete they heard the gunfire.

"Keep going," Fenniman said, with forced calm. The small eruptions stuttered, fading off then erupting again. They could hear nothing else but uncertainly oscillating screeches, which Zac took for screams.

"What's going on?"

"We're under attack, it seems."

"We've got to help."

"Can't until the doors open. We've got seven minutes. Keep going," Fenniman repeated, punctuating with a drawn Walther. He waved it vaguely at Zac, but his eyes were on the monitor.

"I knew he had a mole in this organization," Zac commented.

"Yes, Jean said. Very perceptive."

"You."

"Think so?" Fenniman screwed a silencer into the nose of the Walther. The gunfire had died down, with unnerving, fleeting bursts. On the monitor figures appeared at the far end of the corridor.

Fenniman shot out the cameras, the gun

coughing enough to startle Zac but inaudible beyond the walls. "Video silence," he explained. "Now they can't see us. Do what you came for, you've only five minutes before the door opens."

"They'll get in."

"The first ten won't. After that..."

The men outside wore black hoods opened only at the eyes and nose, and they milled around uncertainly. As a dark figure paced deliberately toward them, each foot rising and dropping like a great and cumbersome weight, the men snapped to military order.

"This is odd," Zac said. "In Golden Boy's lower torso, there's a large coil of wire."

"An electrical coil?"

"No, it's not tight enough. And it's not connected to anything. It doesn't have any apparent function. The rest is so sparse, it's weird."

"Spare parts then. An oversight. It's irrelevant. Did you find out anything?"

The dark figure outside the door turned, and on the monitor, Zac saw the face of the Man of Iron. His face had changed, become sharper and more angular, with none of the humanizing curves. Why not? Zac thought. Snap off the face and replace it to taste, the Man of Iron could do that; his body was not essentially him, but a shell for his integral components, what passed for his personality. In the robot's grip was an agent, one of Severn's clerks, his teeth clenched in pain as the iron hand squeezed his skull.

"Here?" they heard the Man of Iron ask.

The clerk violently shook his head. The iron grip tightened, and the man screamed, to the delight of the others. Dropping to his knees, his legs giving out, he shook his head again. "No one comes here. They're just storage rooms."

"What do they store?"

Three minutes.

"Nothing. Nothing! They've been...going to take them out. Budget problems, there's no money for it."

The Man of Iron tried the door. It held, even against his strength. With the slightest flex, he broke the clerk's nose.

"Open them!"

"I can't!" he sputtered back, blood gushing down his face as he tried to stanch it. "Timelocks, they only open once a day. I don't know when. It was all part of the storage system once. It's the truth!"

In a blink, a red spray blotted half the monitor, and as it dripped away, they saw the Man of Iron, its hand slick, the clerk broken on the floor. The robot beat on the door three times. Zac and Fenniman held their breaths, but the door remained secure. Iron eyes stared at the camera. His fist pistoned up, and their monitor went blank.

One minute.

"We're dead," Zac said. "That door opens, he comes in, and that's the end of it."

"We can't be sure what's going on out there." There was no bravado in

Fenniman's voice now, just a cold despair struggling for hope. He stripped off his jacket and shoveled the pieces of Golden Boy into it. "Take this. I'll go out first, and distract as many as I can. You'll have a fighting chance. Run, don't look back. At the end of the corridor, go left, head down the stairs. All the way. There's an escape tunnel, they built it in case the place was bombed during the war. When you get out, run! Just keep running! Get that thing—" pointing at Golden Boy—"away from here."

They watched the monitor count down. The air seemed too hot to breathe, and Zac noticed tears running down Fenniman's face. Three seconds, two.

"Ready?" Fenniman asked. Zac nodded.

The display came up double zeroes. The door clicked softly, then whirred. Fenniman held the pistol out in front of him, his hands shaking, and as the door swung open he lunged out.

The corridor was empty, dark and still. Zac stepped out cautiously, struggling to carry his bundle. His foot brushed something soft and cold: the clerk's body. His skull was torn away, the brain removed.

"Hurry!" Fenniman hissed, barely louder than a whisper. Zac began to speak, but Fenniman signaled for silence and urged him on, waving the pistol. They hurried to the end of the corridor, but the Man of Iron and his terrorists had left. Fenniman listened at the head of the stairs down, heard nothing, and shoved Zac toward them. He gestured for the agent to follow, but Fenniman shook his head, then turned and sprinted the other way, into the complex. One last time, he stopped and waved Zac downward, then vanished around a corner.

The building was a tomb.

Zac ran down the stairs, taking care not to spill his package.

—◁◠◠▷—

For five hours he stumbled down dark English country roads, taking cover at any approaching sound. He slept in a ditch, the remnants of Golden Boy bundled in his arms. By morning, he saw no signs of pursuit. For the first time in weeks, he didn't feel as if eyes were watching his back. There, in the countryside, he was completely alone and free. If Fenniman hadn't talked, no one would have any idea where Zac might be or what had happened to him.

He only took comfort in that for a moment before his thoughts turned to Jean, and he felt like a coward. The image flooded back that had come to him before, in the room, when the Man of Iron appeared: Jean, stepping in his way, bullets riddling her in a breath, life burbling out of her. He hoped she had escaped or had left before the assault. More than anything, he realized, he wanted her alive.

But there was no turning back.

By mid-morning, he reached a village, and bought shipping boxes and tape at the post office there. Carefully, he separated the parts of Golden Boy into different boxes, a pound (he guessed) in each. The postmaster, a frail but steely old woman, glared at him as he mailed the packages, and on the customs form he wrote "machine parts" when asked to declare contents.

With the last of his money, he bought a lunch of fish and chips, and a train ticket to London. He knew they could be waiting for him there, but he needed his passport. As the train pulled around a curve, he recognized Hembley Manor on a distant hill, and could see police barricades on the road leading up to it. Smoke hung in the air over the manor, and Zac could only guess what had happened. He closed his eyes and slept the rest of the way south, Jean's phantom dead eyes haunting his sleep.

—◁◠◠▷—

He dodged fellow students and sneaked unseen into his dorm room. As far as he could tell, no one had been there. His passport was still in a desk drawer, with his credit cards. At his bank, he maxxed out the cards for cash, and traded in his bank card for the contents of his account, what was left of his scholarship, until he had enough money to get by in his pocket.

He took a cab to Heathrow. In the airport shops, he bought a suitcase and filled it with a handful of clothes and a toiletry kit. He bought a ticket for America and checked the bag. Without it he would have been memorable, now he was just another tourist. At the currency exchange, he traded some British pounds for American dollars and stuffed the rest in his shoe.

At a rack of pay phones, he made one call.

"Hello," a voice answered. Wilcox.

"Hello, Dean. Tell Severn I'm all right."

"Who is this?"

"You know who."

"Robillard?"

"No names. Just pass the message on."

"I'm sorry, I don't know who—"

"Look, you —" Zac calmed himself. "I don't know his number, I know yours. I don't know how to reach him, you do. Don't pretend otherwise. I'll call him in a week, where you are. Tell him."

"Robillard —"

Zac hung up. His plane was boarding.

America

Customs at O'Hare waved him through, to Zac's relief. He feared having to explain the nearly £2000 in his shoe, but after eight hours of flight, all the passengers looked as seedy as he felt. He traded in another handful of pounds, collected his bag—he was looking forward to changing clothes—and rented a car. It was a risk, showing his drivers license, but what choice did he have? He paid in cash to leave no credit card trail, and even *if* someone stumbled across the car rental, they'd have no way to know where he was going.

He drove north from O'Hare on the Interstate to Wisconsin. At Madison, after a snack, he veered onto a state highway and continued north as the Interstate curved west.

At last, just north of Portage, he arrived at his grandfather's farm.

Old man McAvoy answered the

door, staring at Zac disgustedly. "Something I can do for you?"

"It's me, Mr. McAvoy. Ben's grandson. You bought the farm from Ben, remember?"

At last the man relaxed. "Sure, I remember you. Sorry, thought you was a government man. They're always coming around for some fool thing or another." He spat.

"If anyone comes asking for me, I wasn't here. Okay?"

"You in some kind of trouble?"

"Government," Zac replied, and McAvoy spat again. The word was a talisman, connecting them in a mystical rebellion.

"Coming in?"

"No, I don't want to trouble you. I just need to know if my grandfather left anything here. Books, papers, that sort of thing."

"Nope. Your dad put everything in storage, far as I know. Sure there's nothing I can get for you?"

"You know anyone else who, uh, has a problem with the government? In case I need, um, help…"

"I don't belong to no militias, son. Too old for that."

Zac shook his hand. "Thanks, anyway. I'll be heading along. Remember, I wasn't here."

McAvoy chuckled. "Might know where to find me some militia, though…"

—◦◦◦—

Zac drove southeast to Milwaukee checking for any signs of pursuit, and found the U-Rent lot, 24-hour self-service. He parked the car six blocks away and walked over, watching until nightfall. If anyone else knew about the locker, they weren't making themselves known. He waited until the attendant had settled into his tiny office for the evening, when the TV glowed blue on his face,

and then Zac scaled the fence and crept back to his grandfather's locker. He had three flashlights in his pocket.

He spun the dial on the combination lock as he had dozens of times before, with his father, when he was younger. The door opened. They called them lockers, but it was really a small room, 10 feet by 10 feet, and this "locker" was piled high with boxes. He switched on a flashlight and closed the door behind him, sliding a bit of cardboard between the bolt and frame, to keep the door open. If he was careful and quiet, no one would know he was there.

He began to tear open boxes.

Voices woke him: a woman and her children, at the locker across the way. Noon by his watch. He peeked out. The family walked away, not seeing him, and he slipped out, locking the door and fell into line behind the family. The attendant, a different one, was busy helping a young couple, and paid no attention. Zac ran to his car, filled the tank at a gas station, then drove back to storage lockers and signed in using his father's name. Again, he had no choice but to leave a paper trail, but he hoped that doing all the work before showing himself would buy him time. He loaded the boxes he had found into the trunk and kept driving until he had reached the Missouri border.

—◦◦◦—

He booked into a truckstop motel and slept fitfully. The next morning, he drove west, through Kansas City, before turning south. From a coffee shop in Wichita he placed a call: another risk, but again there was no way around it. A woman's voice came on the line—"Southwest Books"—and Zac knew it.

Amy, briefly a girlfriend in high school. Their mothers had kept in touch, even after Amy's family moved to Wichita, and Zac still traded Christmas cards with her, though he hadn't spoken with her since well before he left for England. Even so, her voice was warm and familiar, unconcerned, and she had never easily hidden her emotions. The two words were enough to tell Zac no one had been in touch with her. She was a random thread in his life, frayed and fallen away, and he hoped time was on his side, that anyone digging into his past would only find her after doubling back over old ground, if they learned about her at all.

"Amy," he said. "Don't say my name. Did you get my packages?"

He could hear the surprise and confusion in her voice. "Why did you send them?"

"I'll tell you when I get there, the day after tomorrow. You can leave the books in your storeroom, that's fine with me. I'll see you then, Amy."

"What?" she said, mystified.

He hung up and looked at the small bookstore across the street to see Amy waiting on a customer, plain as day in the shop window. He ordered a Denver omelette, then took a seat in a window booth, to watch for surveillance before he tried to see her.

As Amy locked the front door of the shop, Zac walked slowly around the block, reading a newspaper to cover his face. A parking alley ran behind the row of stores, and he darted down it, arriving at the back door of the bookstore just as Amy was coming out. She was shocked but not upset to see him, a good sign.

"Zac? You said—!"

"I know, I know. Are my packages here?"

"What's going on?"

"You're better off not knowing, Amy. It's nothing to do with the law, I can promise you that. How many of my packages got here?"

"I don't know. I didn't count. There were boxes and boxes."

"A couple dozen."

"Yes, easily."

"Good, I need them."

"Now?" She sighed before he answered, noting the desperation in his eyes, and unlocked the door.

"We'll load them in your car and move them to mine."

"I have robots to do the lifting."

"No. I'll do it. Robots log activities, and I don't want to leave a record."

"Are you sure this isn't anything illegal?"

"Just dangerous," he said, as he brought the first set of boxes out. "Thank you, Amy. Thank you."

She smiled nervously.

⋙⋘

He curved south, heading toward Texas and took a motel room in Lawton, Oklahoma. Out of the way, he decided, as good a place to stop as any. The car was weighted down with boxes, and the week he had rented it for was nearly up. Soon, it would come up missing, and they would likely list it as stolen, making him a target. He had to learn what he needed by then.

Besides, he was out of places to go.

In the middle of the night, when

no one was watching, he shifted the boxes into his room. Cars cruised the strip mall across the street, but no lights hit him. He still had creeping paranoia, but at last he understood no one was on his trail. No one had any idea where he was, and that meant he could breathe again.

With "Do Not Disturb" hanging from the doorknob, he spread the contents of the boxes on the floor. Excitement burned away his fatigue, adrenalin pulsed through him. Carefully, he organized Golden Boy's inner workings as he remembered them, having discarded the shell in the English countryside. He sorted his grandfather's notes and papers, and a stack involving the robotics project rose before him. He organized the papers by date, and when that was done, he slept.

Tomorrow, he would read.

⊶⚬⚭⚬⊷

The drawing, an electrical diagram, was not in his grandfather's hand. Creed, Zac assumed. It disconcerted him how advanced their designs were for the time, and how much things had changed in half a century. During the war, the parts and equipment to turn such a design to reality must have been rare and expensive, and here, he had walked to a Radio Shack to get everything he needed, for next to nothing.

In the papers, far more obviously Benjamin Robillard's than the forgery pushed on Zac in Switzerland, he found most of the answers he sought. Only the loose wire still mystified him. His grandfather hadn't even thought to mention it.

He checked the clock: almost a week exactly, allowing for time zones, since he had called Wilcox. Zac packed everything away for a quick escape if necessary, then walked to the drive-in across the road.

"Is he there?" Zac asked, when Wilcox answered. Zac kept his eyes on his watch.

A skittering of hands over the phone, and he heard Severn's voice, worn and tense. "Robillard, where are you?"

"Nowhere. What happened?"

"You were right about a mole. A security guard named Edom. He brought them right to us. He was after you, Zac."

"Duh. Anyone get hurt?"

"Dozens."

"Fenniman?"

"Dead."

"Jean?"

"That's...more complicated. Let me bring you in, and I'll give you the details then."

"No, thanks. I've experienced your hospitality. Don't worry, Severn, I'll get him for you."

Twenty eight seconds, not enough time for a trace to go through. Zac hung up. He bought a shake and a burger at the drive-in and went back to the room. As he opened the door, he froze, the shake slipping from his fingers to spatter on the threshold.

Severn sat the edge of his bed, a cellular phone in his hands. His face was tired and drawn, his eyes dark and puffy from lack of sleep. He looked, Zac thought, like the walking dead.

"How?" he said.

"What?"

"How will you get him?"

"Where are your men?"

"I'm here alone. I haven't many men left."

"How did you find me?"

"We have satellites made to count the nail clippings of Russian generals. Did you really think we wouldn't be able to keep track of you?"

"So, I guess we're going in."

"No," Severn said unexpectedly. "We're not. What was your plan?"

"Figured I'd go underground, hook up with some militia groups. They're supposed to be linked with neo-Nazis, I get in with them and someone's got to know where the Man of Iron is."

"A bit roundabout, and it would take years. Why not go straight to him?"

"You know where he is?" Severn nodded. "Why don't you go get him?"

Severn stared at the floor. "As I said, it's complicated. He took Jean. He thinks she's important to us. A hostage, for trade."

"For me."

"And the robot. He was very specific. If anyone besides you goes to him, Jean will instantly be killed. I can't risk her, Zac, she's my daughter."

"Listen, I've almost got this robot figured out. I think I can rebuild it, modernize it. It could work for us. Designs are so advanced these days, he'd be almost human, at least in his movements."

"I can't let you jeopardize Jean."

"She's already jeopardized," Zac said. He scooped up the gray sponge and waved it. "Do you have any idea what this is? It's a brain! Grandad grew an artificial brain out of some kind of metallic compound, like silicon chips are "grown." I haven't been able to analyze the material, because I don't have the tools, but Severn, listen! This is brilliant work! It subdivides, the way a human brain does, different areas control different functions. See this?" He showed the circuit board he had spent the afternoon building. "They had a playback system. This brain will play back whatever experiences it has recorded. We can finally learn exactly what happened that day in 1945." He attached alligator clips from the circuit board to the brain, and plugged in a small speaker.

"Play it," Severn said.

◆◇◆◇—◇◆◇◆

They rolled through the dark Texas night, Zac behind the wheel, Severn riding shotgun, staring forlornly at the infinity of stars overhead. "So, with his dying breath, Hecht ordered the robot to bury itself with him. He wanted to keep it out of German hands that badly."

"I think," Zac said, "he wanted to keep it out of everyone's hands."

"Why? We built one, and he knew we knew how to do it. What possible purpose could hiding Golden Boy serve?"

"It's the wire. He told it to hide the wire. Why? I checked it, and it's just wire. But it's not native to the robot's design. It was thrown in after the robot was built."

For miles they rode in silence. Abruptly, Severn said, "1945."

"What?"

"We used wire recorders during the war. Tape has been in use for so long wire doesn't enter anyone's mind anymore."

"It's a recording? Of what?"

"Well, it didn't originate with us."

"So Hecht got it from...?"

"The Germans?"

Zac jerked the car to the highway

shoulder, skidding to a stop. Leaving the motor running, he stepped into the cool night air. Severn followed him from the car.

"Zac?"

"That's what he wants," Zac said. "The wire is his. He needs what's on it. Where can we get a wire recorder?"

"The University Of Texas, perhaps. Failing that, I'll have one flown in from the Smithsonian or the London Museum."

"The university...could I use their robotics facilities?"

"I could arrange that, yes."

"Get in," Zac said. "We're going to Austin."

❧

After a train of shrill squawking, the wire spooled out of the recorder and thwipped against the table. Severn switched off the machine.

"Gibberish," he said.

"Get a tape recorder," Zac said. "Run the wire through again. I want to tape the noise."

"But it's incomprehensible."

"Ever hear a modem over a telephone? Shrieks and scratches, just like that."

"You think it's some kind of data?"

"I think the Man of Iron isn't interested in accessorizing. We get a tape, the computers here should be able to translate it. Just like cracking a code, you look for repetitive patterns. Get enough, they start making sense. Worth a try."

Unconvinced, Severn said, "All right."

❧

"How's it coming?" Severn asked, but Zac didn't answer. He sat in the study room with the lights off, almost lost in the gloom, staring past the monitor playing a looping sequence of screens.

Severn rocked his shoulder. "Are you all right? You look ill."

Zac pointed at the screen and said nothing.

Severn read several screens and staggered back, the color drawn from his face. "That's monstrous!"

"Human brains," Zac muttered. "The best storage system Grosswald could come up with. I'd guess that's what became of that agent back in '38, the one McBain was all bent out of shape over." He took the wire from his bag and twisted it until it was kinked and knotted, useless. He hurled it at Severn.

"Throw that in a furnace," Zac continued. "I don't want anyone to ever see that again." He tapped a sequence on the keyboard, losing the translation to an unrecoverable delete. "I want your word you'll never tell another person about this."

"Agreed," Severn said. "He's a monster. He must be destroyed. I'll call in a strike —"

"No. Not until Jean's safe."

"She's my daughter, Zac! But this—this—I...remember McBane taunting your grandfather with the image of thousands of Iron Majors overrunning Europe. I thought it was a cheap fright, but he knew. On some level, he knew. We can't take a chance..."

"Listen to me! We won't be taking a chance. Have a force on alert, and if I can't stop him, then you can call in a strike. Get rid of that wire. We have things to do. I finish the Golden Boy redesign, we stop at a hardware store and we're out of here."

"A hardware store?"

"I need another piece of wire," Zac said.

B o l i v i a

He stood on the prow of the boat, little more than a raft with an outboard motor and small cabin attached. The captain, a small native who never dropped his idiot smile, sat comfortably at the controls, habitually ignoring the jungle heat. The other passenger, bundled like an eskimo, his face constantly hidden, sat unmoving in the back of the cabin, taking no food or drink, and twice the captain asked Zac if the man had died. They had been sailing along Bolivia's eastern border three days, upriver from Riberalta, where Severn had left them. Now and then he saw shadows pacing them along the riverbank, but the runners slipped from view before he could make them out. Rubberneckers or spies? No matter, they were now the game he played to keep his mind off things.

"Bring her back" were Severn's last words to him, and what could he say to that? It had been more than two weeks since the debacle at Hembley Manor, with no further contact from the Man of Iron. The captain knew the name of the town Zac had given him, a village accessible only by boat via the Guaporé, and was content enough with Severn's money to make the long trip. The lack of event on the voyage only gave Zac more time to doubt himself, and he felt like a kid now, a stupid kid, rushing to trouble. But, he reflected, he had never behaved that way as a boy; perhaps he was making up for lost time.

At last the captain pointed out the village dock in the distance. For the first time, Zac realized he would almost certainly die here, to be as unknown and forgotten as Elliot Hecht had been in Switzerland. He suddenly wanted to run. Had Hecht wanted to run? Had he wanted to live? Zac doubted Hecht ever thought of such things. He was a hero, and heroes did what they had to do, or at least they did in all the books and movies.

To Zac's surprise, no armed men greeted him at the pier. It was empty, despite a number of rafts and fishing boats moored there. He brought his companion ashore, then told the captain to continue sailing and return in six hours. The idiot grin glanced once back at Zac, and the boat was gone. Still, no one came to greet them.

They walked into the village. Aluminum huts, no doubt someone's idea of affordable housing and completely wrong for the stifling climate, dotted what passed for streets. The air stunk of rot, and large green flies were everywhere. He swatted them away, and walked to the end of the village. Grass sloped gently up a ridge, and Zac left his companion at the village edge and went to investigate. The stench grew as he approached, and at the top of the ridge, he gagged. Beneath was a sinkhole, dozens of corpses hurled into it, most wearing armbands marked by a swastika.

What, he wondered, had happened here?

As he turned, he saw a second figure, tall and dark, metal gleaming in the equatorial sun, along side his companion. The Man of Iron. The robot stood and watched, and made no move but to gesture him forward.

"Robillard," he said as Zac neared and stretched out his hand. Zac refused the gesture.

"Where's Jean?"

It seemed as if the Man of Iron were smiling. "Soon," he replied, then looked at Golden Boy. "Why hide him?"

"Golden Boy, take the clothes off."

When the coat and hat fell away and the gloves and scarf stripped off, a lithe form stood there, indistinguishable in shape from a man, resembling a Greek god instead of a machine.

"Impressive," the Man of Iron said. "Perhaps you will do that for me."

"Where's Jean?" Zac said as they walked back into the village.

"Here," the robot said, and tapped his forehead. The meaning escaped Zac at first, and when it hit him, he dropped to his knees and vomited.

"Why?"

"It was a matter of pride. She opposed me, so I used her. I needed…" It stopped, uncertain, and Zac sensed a confusion he hadn't previously seen in the machine. "I'm…"

"You're falling apart," Zac said. "You've been going for fifty years, and you're falling apart."

"It seemed so...elegant... I was... I was human once..."

"You were based on a man once, maybe, but you were never human."

"I was Baron Uwe Grosswald."

"No. You weren't. A man isn't pure data. He can't be recorded. You were a copy, that's all, faded and imperfect. And the only way you stayed intact was to use other brains to keep you going."

"You have the wire."

"Yes."

"Give it to me." Zac handed over a length of wire. "There was a cave-in, I went for too long. I...lost part of my memory. I knew how to... how to keep myself going, but..."

"But you couldn't repair yourself. You didn't remember the technology anymore. But it was on the wire McBane's agent stole from you. Your data, and you needed it back, and when you found out about me, the grandson of the Golden Boy's creator, you thought I could it find for you. What happened to your men?"

"Useless, all useless. I couldn't create my army. Brains and more brains and...I've burned them all out so quickly. They were just men and...I needed power" The Man of Iron staggered into a hut and disappeared, leaving Zac standing there, trembling.

From inside the hut came a grating, animal scream and crashing, and the dark robot staggered out, howling. It saw Zac and charged wildly.

Golden Boy stepped between them. It waited for no command. The two machines collided with an oddly muted clang, iron thudding Golden Boy's new metallicized plastic skin.

Man of Iron stopped, gazing pathetically at Golden Boy, as if the shining robot was unknown to it. "Step away," Zac ordered, but Golden Boy stood its ground, and Zac suddenly knew he no longer controlled the robot. Its arm swung, microprocessor-controlled servos adjusting the punch for maximum force, knocking the Man of Iron back.

Unless, Zac thought, it had silently bided its time, as the Man of Iron had. It was a staggering idea: Golden Boy, with a consciousness formed when? When he rebuilt it in Texas and armed it with state of the art electronics and microchips housing power whole buildings couldn't have contained when Golden Boy was "born?" He thought of Switzerland, when the robot literally rose from the grave to save him. But had it? What had called it up? The family name, invoked as a password, or the metallic-gravel voice of its most hated enemy?

And here they were again, the two behemoths, fighting. Zac watched sadly. The prototypes of the next race—he knew in his heart there would be a next race—locked in war, repeating from the accident of their birth the horrors of their creators. Toe to toe, they struck each other again and again, leaving huge dents and ripped "flesh."

Slowly, Man of Iron gained the advantage. It was unplanned: the robot's mind was clearly destroyed, wiped clean by the overwrite from the blank wire Zac had fashioned, but its ferocity grew. What drove Golden Boy was uncertain, but it lacked the animal fierceness of its opponent, and the cold iron blows tore pieces from its limbs and torso, exposing and damaging electronics, leaking fluids. A break was forming in Man of Iron's chest, its head sat slightly ajar, but still it pushed, implacably: a machine following its final programming.

A leg cracked under Golden Boy, and it tumbled. Man of Iron stood over it, lifting iron hands for the finishing blow. Golden Boy was finished.

Then it happened: Man of Iron stopped where it was, and stretched its arms wide. Its stance changed, all the rage gone, and it tilted back its head, exposing its chin.

Golden Boy punched upward, a blow that

collapsed its own arm on impact. Man of Iron's head flew backward, torn from its body, which convulsed, sparks spewing from its neck, and pitched forward, to lie still in the dirt.

Zac heard a voice calling his name, so softly it might have been a dream. While the voice had the Man of Iron's familiar rasp, his inflections were gone. They'd become feminine.

"Zac," the voice said again. He picked up the fallen iron head, its face half torn away. Floating in a cracked translucent dome inside was a human brain, electrodes gently touching it.

And he knew: the only consciousness left in the head.

"Jean?" he asked.

"I...stopped him..." The voice crackled, fading in and out. The brain was dying.

"We did, Jean. We did. I'm so sorry. How long have you been conscious?"

"They...were all...conscious...they died...twice..." A horrible thought: your last awareness spent in silent slavery to the creature that destroyed you.

"He...got what...he wanted..."

"He didn't. We won."

"He...didn't know...but...he wanted to...die...He wanted...to...feel the rain...on his face... but he...couldn't...you don't know...what it's...like to be trapped...in metal...promise...promise..."

He lifted the head gently, desperate to preserve it, and keep the brain alive. But he knew that was beyond him. "What, Jean?"

"Never...again...do not case...a mind...in metal... Never...you have to...end this...he collected...munitions...a hut at the...south...end of...they thought...he planned to conquer...the world...but other...reasons...he hid from...himself...Destroy...me...all trace...of him..."

"No, Jean. Don't give up. Maybe we can..." He stopped, defeated.

"Free...me..."

Then Golden Boy was beside him, reaching for the head. Zac shoved him away, but Golden Boy gently pushed him aside, snatched up the head and was gone. Zac slumped to the ground as the exhaustion of the last two weeks caught up with him and he succumbed to blackness.

◁◈▷ ◁◈▷

A terrible noise slapped him awake, a roar like a raging beast bigger than the world. A pillar of flame jutted from the eastern end of town. There was no sign of Golden Boy or the Man of Iron's head or body, but from the flames rose the distinctive sound of plastic fragmenting in the fire.

The blaze spread quickly from hut to hut, racing for him. Zac shook himself alert and ran to the dock and hastily tore a raft free. On the river, drifting to midwater, he paused to look at the village. It was already gone, and the jungle burned, and the grasslands that lead out to the sinkhole. The inferno would scorch clean this small world.

Bits of burning plastic ash rained from the sky, fizzling to black goo in the water.

"Good-bye, Jean," Zac said. Her last words echoed in his head, also echoing thoughts that had struck him days earlier, while reading his grandfather's papers: a race of robots, not metal, not constructed, but made of flesh, grown as Benjamin Robillard had grown the artificial brain. Artificial intelligence had been unleashed, and genies never went back into their bottles. But metal was already the past. There would be new technologies, and better, if he had to invent them himself.

He would open the way, but she would always be his
muse. It was all he could do for her, now that she was dead.
He cried uncontrollably, and floated downstream.

THE END

PROTOTYPE

Written by
John Gregory Betancourt

Illustrated by
Mark Jackson

The Vienna Express
September 15, 1936

As he struck a match and leaned back to light his Lucky Strike, Flynn took a surreptitious glance around Baron Ogilvy's private car. Red velvet upholstery, gilded wood and highly polished mahogany filled the room. The dozen or so noble-born Austrians sitting in the high-backed chairs chatted amiably in German and French; a few others worked at opening more bottles of champagne. *Poor fools*, Flynn thought. *They hadn't a clue about what lay ahead.*

He took a deep drag on the cigarette, then slowly blew out a cloud of light, gray smoke. His acting coaches had taught him to use props like this cigarette so he always looked natural. He took another pull and tried desperately to keep his tension from showing. *We're almost there*, he told himself. *In an hour, when we cross into Austria, I'll be safe.*

The danger didn't come from the baron or his retinue; they were the cream of Viennese society, stunningly clueless men more interested in hunting than politics, accompanied by beautiful women all dressed in the latest Paris fashions, lace and diamonds and pearls at their throats and ears and wrists. The baron and his friends circled through the European social scene, drinking, hunting and celebrating a lifestyle which, Flynn knew, would soon end with the coming of war.

He had met Baron Ogilvy in Berlin at the grand opening of his latest MGM picture—*They Dared the Alps*—which paired him with Katherine Hepburn and Van Johnson. Kate and Van had the easy part, touring the United States to promote the movie. Louis B. Mayer had steel-armed *him* into the European tour because he spoke passable French and German.

The baron, learning of his plans to promote the film in Austria next, had invited him to share his private car aboard the Vienna Express. Flynn had agreed immediately.

Now, he sagged deeper into his seat and half closed his eyes, trying to look as though he were drowsing off. Maybe they would leave him alone if he did. He'd done enough today. Enough that it might cost him his life….

"Mr. O'Conner," a beautiful young woman said to him in lightly accented English. She touched his arm, held out a champagne glass and, with a smile, he stood and accepted it.

"*Danke schoen*," he said, sipping. It was French, of course, a light, dry champagne, and quite good.

"Please, Mr. O'Conner, my English is excellent."

He forced a laugh. "Better than my German, I'm sure. Will you join me?" He indicated the seat beside his. Courtesy required no less.

"Certainly."

With a laugh, she sat, and he sat next to her.

"Aren't you Baron Ogilvy's niece?" he asked.

"Yes. My name is Gertrude Berliner Ogilvy. My friends call me Gerty." She winked at him, then traced a pattern on his leg with her fingernail. "I'd like it if you called me Gerty…Flynn."

Flynn felt a sudden hot flash. She was making a pass at him, and right here, in front of the baron and everyone.

Before he could reply, the door at the far end of the car suddenly swung open. An attendant in a white uniform and cap stuck his head in. "Border in five minutes, sir!" he said to the baron.

Flynn suddenly felt light-headed. This was it. They would be waiting for him here. He groaned inwardly. How did he get himself into this mess?

The meeting in the War Office came back to him.

"I'm an actor, not a spy," he'd said to Ian McBane. "Uncle Sam doesn't need me."

"That is *precisely* where you're wrong," McBane said. "Rumors of major war are spreading quickly in Europe. When it starts, it's only a matter of time before the United States becomes involved. Remember what happened in the Great War, after all. They needed us to settle things."

Flynn nodded. His father had been wounded fighting the Kaiser.

"It won't be risky," McBane told him. "You're a celebrity. No one will search you at the border. All you have to do is carry some papers out for us."

That had seemed simple enough, so he had agreed. But then his contact had shown up with a bullet wound in his chest, and before staggering off, the man's last words had been: "*The Nazis are planning to invade Austria. Guard these plans with your life. You must get them home safely!*"

The news had stunned him. It's one thing to say, "War is coming," and quite another to say, "The Germans are invading in a few months." The Nazis—testing new weapons and warming up for battle in the Spanish Civil War—would rip through Austria unop-

posed if he didn't act. Austrians, like the baron and his friends, suspected nothing…they all thought Hitler would be happy to flex his military muscles in Spain.

Flynn had taken the papers that morning. Ever since then, he'd felt the eyes of the Nazi secret service upon him. In the lobby of his hotel, in the press meetings, at the afternoon reception he attended, men in dark suits had watched his every move from the shadows. They hadn't arrested him, he thought, because he was a celebrity and kept himself surrounded by crowds.

The baron's private train car offered the perfect chance for escape. If the Nazis intended to arrest him, they would have to do it here, in front of the baron. If not, he would be safe in Austria in a matter of minutes.

He touched his suit's inner pocket. He still had the papers. If they *did* search him …

He swallowed and suddenly felt sick. No, he had best get rid of them, at least until they made it past the border. Where could he put them? He glanced up the length of the baron's car.

"Are you well, Flynn?"

"Pardon, but I'm afraid supper did not agree with me." He rose, nodding politely to her, and headed for the tiny washroom at the back of the car.

Thankfully, it was empty. He went into the cramped little room—sink with broad makeup counter, small toilet, spittoon on the floor—and bolted the door behind him. Sagging a little, he stared into the mirror: steely blue eyes, sharp high cheekbones, slightly flared nose, long blond hair falling over one eye, skin pallid as death. He looked a wreck.

He shrugged off his coat, hung it on a hook and rolled up his sleeves. After plugging the sink basin, he poured a couple of inches of water from the pitcher and splashed cold water on his face. There—marginally better, he thought. He dried off on one of the baron's gold-monogrammed towels, then checked his pocket watch. Nearly eleven o'clock. He felt sweat beginning to trickle under his arms and down the small of his back.

Brakes squealing, the train began to slow. Flynn drew in a deep breath; best to get it over with, he thought, shrugging on his coat and straightening his tie. A movie star has to keep up appearances, he told himself, forcing a stage smile. He combed back his hair neatly. There—he almost looked normal again.

Next, he pulled out the six thin pages of war plans, folded them up as small as he could and wedged them up under the washbasin's counter, between the cabinet

and the wall. You couldn't see them unless you bent over and looked. That would do for now, he thought. He would reclaim them as soon as he made it to Austria.

He went back into the baron's car. Everyone crowded to the windows, looking out.

Flynn joined them, leaning over Gerty's shoulder. Her hair smelled of lavender, he found. It might be nice to spend the night with her, once they made it into Austria. *If* they made it.

The border station was quite small, little more than a depot lit by the yellow glow of gas lamps. Mountains rose in the background, though this late at night they were little more than indistinct shapes. Perhaps two dozen German guards in uniform stood rigidly at attention on the platform.

The train drew to a stop. Two German officers began to pace the length of the train, shouting, "Everyone off!" in German, then bad French, then worse English. "Papers ready!"

One of the baron's men opened a window. "Do you know whose car this is?" he demanded.

"I don't care if it's Himmler's car!" the officer replied. "Our orders are to empty all trains and check all passports on the platform. The sooner you cooperate, the sooner you will continue your journey!"

"Do as he says," Baron Ogilvy said. "It is a small inconvenience, nothing more."

With a few grumbles, the baron's retinue filed off, pulling out wallets and passports. Most of them still held champagne glasses, Flynn noticed. He drew his own passport and fingered the brown leather absently. This would be the last test, he thought.

He lined up with the others on the platform. He could see the officers going down the line, checking documents, stamping passports. A German lieutenant reached him and held out his hand.

Without hesitation, Flynn handed over his passport.

"American," the officer said, studying the papers. In lightly accented English, he went on, "This name is familiar to me. Why is this so?"

"Perhaps you've seen one of my movies," Flynn said, forcing a grin. "I'm an actor."

"Ah, yes?" The officer peered more closely at his face. "You were in *Gunga Din*, ja?"

"That's right. And quite a few others."

"Very good cinema." He stamped the passport, handed it back and moved down the line.

Flynn started to breathe more easily. By God, he was

going to make it! He couldn't believe his luck.

One thing was certain, he'd never agree to spy for anyone, ever again. His nerves just couldn't take it.

The officers motioned everyone back aboard the train. Flynn joined the queue. The baron boarded first.

"I hate these border checks," Gerty said, turning to him. "They are so *bothersome.*"

"I agree entirely," Flynn said.

"Do you have such problems in America?"

"No." He grinned at her. "I don't think we'd stand for it."

Suddenly, Flynn felt a tap on his arm. He glanced over, a sudden panic rising inside.

It was the German officer who had stamped his passport. "If I may," he said, smiling and holding out a piece of paper. "My son likes the American films. Would you sign for him?"

Flynn's grin felt frozen in place. "Of course." It seemed a harmless enough request. "What's his name?"

"Rudolph."

Taking the paper, he pulled out his fountain pen, removed the cap and quickly wrote: "To Rudolph, my biggest fan in Germany, with all best wishes, Sincerely, Flynn O'Conner." He signed his name with an exaggerated flourish.

"There," he said, handing it back.

"I am deeply grateful." The officer blew on the page to dry the ink, then tucked it into his breast pocket.

"Always glad to help a fan." Flynn glanced at the train. Gerty and most of the other passengers had boarded. He reached for the handrail.

"If I may." The lieutenant caught his arm.

"I'd love to sign another autograph," Flynn said, trying to pull away, "but my friends are waiting —"

The officer tightened his grip. The last of the passengers had boarded the other cars. The conductor was staring at them.

"*Gehen Sie,*" the lieutenant said to the conductor.

"*Aber—*"

"*Schnell!*"

Flynn fought a mounting sense of doom. Whirling, he broke the officer's grip and scrambled for the train.

"*Halten Sie!*" the officer cried.

Soldiers rushed from all directions. Panicked, Flynn stopped short. They had rifles leveled at him, and they didn't look friendly. Slowly, he raised his hands.

The train began to pull out. Flynn stared at it, at the windows of the baron's car, where he could see champagne glasses raised in a toast. Gerty and the others hadn't missed him, yet. They hadn't seen the lieutenant detain him.

"This way, Herr Flynn," the lieutenant said.

A covered truck slowly backed up to the edge of the platform, and Flynn found himself forced into the back at rifle point. The lieutenant sat opposite him, no longer smiling. Guards sat surrounding him.

"What is this about?" Flynn tried to bluster. Good thing he hid the plans on the train, he thought.

"We do not suffer spies in the Third Reich," the officer said grimly. "We know very well what you have been up to, Herr O'Conner."

Flynn felt the bottom fall out of his stomach.

>o<

In America and throughout the free world, the disappearance of Flynn O'Conner made the newsreels for two weeks running. Where had the flamboyant actor gone this time? Rumors spoke of secret trysts with Marlene Dietrich in the Swiss Alps or on the French Riviera. O'Conner, the heir to a Texas oil fortune, extravagant playboy, college football hero and movie star, had been known to vanish on drinking and partying binges from time to time. Studio heads felt certain he would turn up…probably with an apologetic grin and a new girlfriend in tow.

Castle Grosswald, Germany
December 23, 1936

Dr. Uwe Grosswald barely noticed the light snow falling outside the windows of the laboratory in which he worked. He focused his attention on the huge black-and-red robot on a table before him. This was it, he thought, the culmination of all his research. The time had come to show his Nazi masters what their fifteen million marks had bought.

"How are the connections?" he asked.

"Perfect," his chief assistant Heinrich Muller said, pulling nervously at his thin mustache with one hand. "I personally checked them this morning."

"Good." Grosswald stepped back and took a deep breath, surveying the line of computing machines that filled one huge wall of the laboratory. They hummed softly, ready to begin sending instructions to his creation. First, though, came the most impressive part of the demonstration.

He glanced at the half dozen high-ranking military officers assembled before him: four colonels and two generals. They all had slightly bored expressions, but that would soon change.

He nodded to Muller, who activated the huge robot's primary power coil. The machine's red eyes lit up; using miniature cameras, it could see everything before it.

"Behold!" Grosswald cried. He drew a small radio-control device from his lab coat pocket. "Metal life!"

As he adjusted the controls, the robot slowly and ponderously sat up, swung its legs off the operating table and stood. It towered over everyone in the room. It had the strength of ten men in its hydraulic-powered arms. The colonels gasped in surprise. Both generals took a step back in alarm.

"Behold," Grosswald said, "the soldier of the future!"

"How does it work?" Colonel Machen asked.

"Using this device," Grosswald said, indicating his remote control, "I can make it stand, walk, turn and lift its arms." He demonstrated, and the robot circled the room on cue, black metal feet thudding heavily on the concrete floor. "It can lift over a thousand pounds and withstand a barrage of bullets...or worse."

"Grenades?" General Heimann asked.

"Grenades, flame throwers, poison gas—anything short of a direct hit from a mortar shell."

"Then it is armored?"

"From head to toe. Mechanically, it is perfect. It only needs a mind of its own to be complete."

"The possibilities for use in wartime are endless," Heimann mused. "It could be as revolutionary as the tank. Why, with a hundred of these metal warriors, I could sweep through any army I face!"

"This is an early prototype," Grosswald said with a dismissive gesture. "It is far from perfect."

"What do you mean?" Heimann demanded. "I see no problems with it."

"This," Grosswald said, holding up the remote control, "is our stumbling block. The robot's operator must stand within fifty feet of the machine. Further, radio waves can be jammed. What good is a metal man who cannot move?"

Heimann frowned. "I see your point, Herr Doktor."

"I intend to make the robot self-thinking. I have in my laboratory the most powerful computer in all of Germany. Watch what an *electronic* brain can do!"

At his nod, Heinrich opened the robot's back plate and ran thick cables between it and the giant computer against the far wall, connecting them. Again, the robot's eyes glowed red. This part of the test was ready to begin.

He spoke into a microphone. "Robot!" he cried. "Raise your arms!"

Vacuum tubes began to glow inside the huge computer. Punch-cards flickered through slots. Gears whirled.

As the computer deciphered his command, the robot began to move. Like a clock, its arms inched upward a tick at a time.

"Why is it moving so slowly?" the general demanded.

"The human brain thinks faster than any computer. It must break down my words into electronic impulses, interpret them, create a program to make the robot move and then execute the program."

Heimann snorted. "Useless. We cannot have cables trailing all over a battlefield, anyway. It's impossible!"

"Within ten years," Grosswald said, "the computer will be small enough to fit *inside* the robot's chest."

"How is this possible?"

"I have heard of research along these lines being conducted by General Energy in the United States, but I have no access to it—it is top-secret work for G.E., too. But the rumors I hear are of a new miniature device called a 'transistor,' which will replace vacuum tubes."

"Why not continue to use a human brain?" another of the colonels asked.

He meant have a human operate the robot by remote control, Grosswald saw, but the suggestion sparked a new idea in his mind. The robot needed the equivalent of a human brain...but what if a human brain could be installed *inside* the robot?

His thoughts raced ahead to the elegant simplicity of a human mind operating a robot by thought waves alone. There were so many similarities between the electrical activity in a living brain and the electrical activity in a computer—and if a computer could run his robot, why not a human brain, too? It should be possible.

"That will be it for today," he said suddenly. He unplugged the robot and watched it sag back, lifeless and inert. "I have important work to do."

"Herr Doktor—" Heimann began.

"I have had an inspiration," Grosswald said. "I will call you again next month. I will have new developments—a fully functioning robot, with no wires or remote control!"

"Is this possible?"

"Yes!" Grosswald breathed. "It can work! It *will* work!"

"If so," Heimann said slowly, "this will be the single greatest advancement in warfare in the twentieth century. Requisition whatever you need." He glanced at the other officers, who nodded their agreement. "We will support your research with the Führer."

Grosswald nodded curtly and turned back to his work.

Vehrnicht Prison, Germany
January 22, 1937

Flynn O'Conner ached. When he tried to stretch kinks from his muscles, the heavy iron shackles connecting his wrists to the stone wall behind him rattled faintly. Burns covered his forearms. Two fingers—broken, then badly reset after his first week of questioning—throbbed dully.

But he hadn't talked. He clung to that one bit of knowledge like a lifeline. He hadn't revealed his part in the plot. The Nazis must still wonder if the papers had made it out of Germany.

The only question that remained was...would anyone find them in Baron Ogilvy's washroom? Perhaps he had hidden them too well. Perhaps he should have given them to Gerty to hold for him. He sighed inwardly. No, it might have embroiled her in this mess. What *else* could he have done? What else *should* he have done?

He kept reliving those last few minutes of freedom in his mind, wondering what he could have done to protect himself or, short of that, to make sure the German invasion plans made it safely to the U.S. Embassy in Austria.

The months of interrogation...the starvation, the beatings, the torture...it would all be for nothing if the plans didn't make their way to proper hands.

Keys jangled outside, then the door to his cell slowly opened. Through swollen eyes, Flynn glared up at the man silhouetted in the doorway.

"Up!" the guard said in German.

"I have told you everything," Flynn said.

"The interrogation is over."

He gaped. "*Over?*"

"*Ja.* You are being taken away from here. On your feet!"

Flynn looked down at his battered, starved body. They couldn't release him looking like this, could they? It would be an international scandal. No, they were going to execute him, he thought. After months of torture he would have thought it a welcome release, but he realized he didn't want to die. He wanted escape... and revenge.

He raised his head defiantly. Well, if he had to face a firing squad, he would die knowing he hadn't betrayed his country. That would be one small comfort.

The guard stalked forward and unfastened Flynn's shackles. Then he dumped out a small bag of clothing... shoes, socks, a light jacket.

"Put them on."

Flynn massaged his aching wrists. Curious...why would they give him shoes and a jacket if they were going to execute him? Or could this be another elaborate torture? A promise of freedom which would be quickly yanked away?

He steeled himself to the inevitable, then with trembling hands began to pull on the socks and shoes, then the jacket. Rising, he staggered from the cell with all the dignity he could muster. Not much, he thought grimly.

When they left the building, he blinked in the sudden brightness of daylight. An inch of snow covered the ground, and an icy wind made him shiver.

Instead of a firing squad, the guard led him to a covered truck much like the one that had brought him here. A few other prisoners were being loaded aboard. At his guard's prodding, he climbed in and sat on the wooden bench. He had survived. The interrogation had ended, and he was still alive. He had won.

The guards climbed in and raised the clapboard, and the truck pulled out. As they drove, Flynn gazed out at snow-covered fields, barren orchards and distant mountains. The guards made no effort to tie the rear of the truck closed with canvas flaps, so he figured it didn't matter if he knew where they went.

As the hours wore on, they stopped several times for fuel and to stretch and relieve themselves. The guards provided cheese, crusty brown bread and watery red wine for dinner. Flynn ate voraciously. Still, the journey continued.

They seemed to be heading into the Rhine area of Germany, he thought. He began seeing castles on distant hilltops. Night fell and still they drove.

They finally turned off on a steep, winding road, full of switch-backs and hairpin turns. They ascended rapidly.

When the truck finally stopped, the guards marshaled everyone out. They were at a large castle. Flynn stared up at high stone battlements and watchtowers equipped with machine guns. It was like something out of one of his friend Errol's adventure movies.

The guards made them line up. Flynn stood shivering beside the other prisoners as papers were signed. The usual German efficiency, he thought; starved, beaten and half dead though they were, they all had to be counted and signed for.

"*Bitte*," a young man in a white coat said. He had round wire-rim glasses, a small mustache and a clipboard. "*Kommen Sie mit mir.*" He indicated a door into one of the keep's larger buildings. Light blazed from the high, narrow windows. At least it looked warm inside, Flynn thought, shuffling forward with the others.

"My name is Heinrich Muller," the man in the lab coat went on in German. "You are to be our…guests."

"What is this place?" Flynn asked him. "Why are we here?"

The young man said nothing but led the way to the door, opened it and held it for them. The soldiers prodded them forward, and Flynn found himself in a large room that might have once been a feudal lord's banquet hall. Now, it seemed to be a hospital of some kind. Medical equipment sat everywhere—examination tables, surgical tools and even an operating theater off to one side. Flynn felt relief. They were going to get treatment for their wounds. Maybe they would even reset his broken fingers so they could heal properly.

Muller then escorted them to a cell with thick steel bars on all sides, even the floor. Luckily, there were benches welded to the back wall. Exhausted, Flynn sat heavily.

A middle-aged man with a round, moon-like face approached, surrounded by younger assistants. Muller joined him, and the two conversed in low voices, looking over the prisoners. Flynn suddenly felt like a slab of meat in a butcher's window.

"Well, well," the moon-faced man said loudly in German, smiling and striding forward. "I see you have all safely arrived. I am Dr. Uwe Grosswald, and you are all here to help the glorious German cause."

Everyone stared blankly at him. Flynn met the man's gaze for a second and saw the gleam of madness there.

"You," Grosswald said, pointing to the man next to Flynn. "What is your name?"

"Adolph Schmidt."

"Come, Adolph, we must get you cleaned up. How long since you have eaten?"

"A few hours."

"And before that?"

"Three days." Adolph struggled to his feet and moved forward. Flynn felt a pang of envy. Aside from what the guards had given them, he hadn't eaten in four or five days.

One of Grosswald's assistants let Adolph out of the cell. Grosswald took the man's arm and led him to one side.

Flynn pulled himself to his feet and pressed against the bars to see. The other prisoners joined him.

The assistants stripped Adolph, scrubbed him down with what smelled like antiseptic, then shaved all the hair on his head. Then they led him to a long metal table at the far side of the room, where Grosswald listened to his heart and began testing his reflexes.

"Excellent," Flynn heard the doctor say several times.

Finally, as Flynn watched, they strapped Adolph to a steel table, took a large bone saw and neatly opened the man's skull. Blood fountained across the room, spattering everything.

Flynn gaped. Bile rose in his throat, but he managed to choke it down. Around him, he heard vomiting. The others hadn't been so strong

Grosswald moved quickly, severing Adolph's spinal cord, and removed his brain with a huge pair of forceps. He handed them to Muller, who hustled the organ into a nearby tank of bubbling yellow liquid.

"Quickly!" Grosswald called, hurrying to another table.

Flynn couldn't quite see what was going on there, but the others scrambled about, wheeling the brain over. Grosswald worked like a maniac, cursing, screaming orders. He seemed to be transplanting the brain into another body, Flynn thought. It was monstrous.

Flynn staggered back to the bench. His vision blurred. He felt a pounding like sledgehammers in his head. *Monstrous.* Monstrous and insane.

Grosswald and his assistants continued to work late into the night. Every time Flynn looked up, they were hunched over their patient.

After six hours, the doctor suddenly threw down his instruments and stalked from the operating theater. The transplant must have failed, Flynn realized.

He looked grimly at his remaining companions. Who would be next?

>∞<

Each day for the rest of the week, Flynn watched as another of his companions was selected, dragged kicking and screaming from the cell by soldiers, then subjected to the same process: disinfection, examination by Grosswald, then death.

It was like something out of a horror movie, Flynn realized, casting the picture in his mind. Peter Lorre would play Grosswald with sadistic glee. Bela Lugosi would be Muller—a cringing, subservient Muller. And Boris Karloff would play the transplant victim. In the movie, of course, Grosswald's creature *would* live.

The real doctor had no such luck. He went through his first six victims without success. Each time the experiment failed, Grosswald stomped out in a fit of anger. Perhaps, Flynn prayed, the doctor would give up before his seventh try, but he knew deep inside that such hopes were useless. Instead of a firing squad, he would die at the hands of a mad scientist. That's why the Nazis had given him a reprieve.

Swallowing, he touched his neck. At least it would be quick, he thought.

But the Texan inside him wouldn't say die quite so easily. He had been bred and raised to fight, and the same stubborn streak that wouldn't let him break under torture wouldn't let him die now without a struggle. There had to be a way out, he thought.

He lay back on the bench, staring up through the bars, thinking. He had one advantage over the others: six full days of rest, with plenty to eat and drink. His strength and reflexes were returning. What other skills did he have? Just acting…

And that, he thought, as a plan started to form in the back of his mind, might just be enough.

✂

Flynn woke as the first light of day filtered down through the room's high, narrow windows. Already he could see Grosswald's assistants sterilizing medical equipment in preparation for murdering him.

He looked down at the scars on his hands and arms. Four months of torture to make him talk, and he hadn't said a word about his spying mission. Those horrors seemed almost clean compared to the fate which now awaited him.

Through most of the night, he had been thinking about how Grosswald had murdered his first victim. Adolph had not been restrained through most of the examination; he had not struggled or fought because he did not know what awaited him. If anyone could have tried to make a break for it, Adolph was the one. *And all because he didn't try to escape.*

That, Flynn told himself, had to be the key. He had to

make them think he'd given up… that he was cooperating. Then, he would seize the first opportunity for escape which presented itself.

He knelt before the bench and folded his hands. The role of Father David in *Chapel Belles* had been his best religious training. He could do pious with the best of them now.

"Please, God," he said so Grosswald's assistant could hear, "If it is Your will for me to die here, I trust it will not be in vain."

Then he heard Grosswald's booming voice calling to his assistants. The mad doctor had arrived.

Making the sign of the cross, Flynn stood and moved to the cell door with his hands clasped and his head bowed. The meek shall escape, he told himself.

Grosswald sent the soldiers to get him. When they opened the cell door, he walked out calmly. Turning, he headed for Grosswald's assistants, who were waiting with brushes and buckets of disinfectant. None of the soldiers touched him. As long as he cooperated, he realized, they saw no reason to interfere.

Calmly, he stripped, folding and stacking his torn, dirty, old clothes to one side, and then he stood silently and let the indignities happen. He pressed his eyes shut and bit his lips when the scrubbing grew too rough and abraded his skin.

Then, reeking of disinfectant, he let them shave his head. Finally, clean and sanitary, ready to be butchered, he walked to the examination table. Grosswald waited there, rubbing his hands together with almost sadistic glee.

As he prepared to lever himself up onto the examination table, the doctor said in German, "What is your name?"

"Flynn O'Conner."

"I want to know…why don't you fight, like the other prisoners did?"

"What is the point?" Flynn replied. "I could not win. In the end, you will have your way."

Grosswald nodded slowly. "Very astute. Your accent…you are English?"

"American."

"You seem familiar to me."

"I've appeared in fifteen movies. Perhaps you have seen some." Could Grosswald be a fan? Could he win his release that way?

"Ah, yes, the cinema, that must be it." He motioned Flynn up. "On the table, please."

So much for that idea, Flynn thought. He turned to pull himself up onto the examination table—and then in one quick motion snatched a scalpel from a nearby tray. Like a tiger, he sprang on Grosswald, pressing the razor-like blade against the doctor's pudgy neck, pushing just hard enough to draw a bead of blood.

"Back!" Flynn roared to the soldiers and Grosswald's startled assistants. "Back, or I'll slit his throat!"

"Do as he says!" Grosswald called in a strangled voice.

Everyone began to back up. The soldiers leveled their guns at him. He saw murder in their eyes. Somehow, he no longer cared.

Keeping Grosswald's body between himself and the soldiers, Flynn began to back toward the door. Reaching behind him, he opened it, and the freezing wind that struck him like a whip brought a shudder to his naked body. He'd worry about pneumonia later, he thought.

Still holding Grosswald, he edged out and looked desperately around the courtyard. A light snow was falling; if he could make it outside the castle in a truck or car, he might be able to lose his pursuers, he thought.

A few vehicles were parked on the other side of the courtyard. Still grasping Grosswald, he edged backwards toward them. Grosswald began to wheeze, but Flynn paid no mind. Of more concern were the soldiers pouring out of the laboratory, rifles raised. The moment he let go of his hostage, he thought, they'd open fire.

He reached the first car, scraped snow from the passenger side window with his elbow and peered inside. It had keys in the ignition, he saw with a mental cry of triumph. This was it—his ticket out of here.

Grosswald began to sag. He'd fainted, Flynn realized. He hesitated. He couldn't manage that much dead weight, not in his current half-starved condition. And he was beginning to shiver uncontrollably from the cold—it had to be twenty degrees out here.

Finally, he opened the door, shoved Grosswald away and dove inside. As he did, something sharp stabbed his right calf. He screamed in pain.

Pulling himself inside the truck's cab, he slammed the door shut. Rifles began to pop. The glass next to his head splintered. There was a hypodermic needle stuck in his calf, he saw with despair. Grosswald had stabbed him with it, injecting him with something. Suddenly, his leg went numb.

No time to think about it, now, though. He stomped on the clutch with his left foot and gunned the ignition. The numbness was spreading. He could barely feel anything below his waist, now. Suddenly, he felt light-headed. If he didn't do something fast, he'd never make it.

He managed to put the car in reverse and began backing up. But when he tried to turn the wheel, he found he couldn't move his arms. He was paralyzed.

"No —" he whispered, the horror of it all catching up. He'd failed.

The car struck one of the castle walls and stopped, wheels spinning helplessly in the snow.

Grosswald opened the driver's side door, reached in and calmly turned off the ignition. The motor growled to a stop.

Flynn found he still held the scalpel. With every last bit of his strength, he raised his hand to stab the doctor.

Grosswald calmly caught his arm and removed the instrument. Then he began to smile. And then he began to laugh like this was the greatest joke he'd ever seen.

"Marvelous," he cried, dabbing at the tiny wound on his neck with a white handkerchief. "Just what we need, a man who's willing to fight to live. Marvelous!"

Flynn moaned. Then everything went black.

Flynn struggled up through dreams of suffocation. He felt drugged, some part of him realized, and that meant he still lived. Grosswald must have put him back in his cell to recover...only a reprieve, but while he lived, he had hope. Next time, he *would* escape.

He opened his eyes and found himself staring up at the ceiling. He was on one of Grosswald's medical tables, he guessed. He tried to roll over, but couldn't move. Strapped down? Still drugged? He had no way of knowing.

His vision was strange; everything looked black and white and oddly flat. Probably an effect of the drug, he decided, the same way everything sounded tinny and muted, as though coming through a lousy speaker system. He struggled to sit up. Grosswald suddenly appeared in his field of vision, leaning over him. The doctor smiled like a kindly uncle offering a child candy.

"Excellent," he said.

Flynn tried to speak, but only a strange buzzing sound emerged. He raised one hand—and abruptly saw it wasn't human anymore. It was black and metal. Armor? It had to be.

"Stand up," Grosswald said.

He tried to grab the doctor but fell to the side and landed on the floor in a clattering heap. He could see the rest of his body now—legs, stomach, torso, arms, hands, all shiny black metal with red highlights.

Slowly, he raised his head and caught sight of his reflection in the side of a steel cart. He was encased in metal. His eyes glowed red, tiny dots like burning embers.

"Stand up!" Grosswald cried. "On your feet!"

Slowly, Flynn obeyed. Why would Grosswald put him in a suit of armor? It made no sense. Not after what had happened to the others ...

Then the full horror of it hit him. He realized what must have happened. Grosswald put his brain in this suit—it wasn't armor but a machine. Suddenly, it all made a sick sort of sense.

Enraged, he turned toward Grosswald, raising his arms. His limbs obeyed more slowly than he was used to, and he seemed to have lost his sense of balance, but he managed to do it. He'd crush the life out of Grosswald, he thought. And then he'd rip this place apart.

"Stop!" Grosswald shouted. "Lower your arms!"

Flynn found himself obeying, despite his every attempt to attack. His body refused to cooperate.

"You've done it, Herr Doktor!" Muller cried.

Grosswald was smiling, "Excellent," he kept murmuring.

Flynn tried to scream.

Flynn spent the rest of the day obeying Grosswald's every command. Stand, sit, fetch, turn in a circle—whatever Grosswald ordered him to do, Flynn found himself doing it. It was a nightmare. He had no power to resist. It was as if the doctor's will took precedence over his own.

That night, they ordered him into a corner before locking up the laboratory. Flynn stood there until they left. As soon as he heard the latch click and knew he was alone, he discovered his will was his own again. With no one to order him to do something else, he could follow his own counsel.

He knew he had to warn the outside world. What the Nazis had done to him was too terrible to ever be allowed to happen again. Grosswald *had* to be stopped before he could build more machines like Flynn.

If only he had better command of his body, he thought. He raised one hand. His every movement was a struggle. He felt like a drunk trying to play pin-the-tail-on-the-donkey.

He couldn't get his body to move with the precision his mind demanded.

Curiously, he didn't feel the slightest bit sleepy. Perhaps that was the advantage of a metal body, he thought. He would never get tired or sick or hungry or cold, just as he would never smell flowers, drink champagne or make love again. All he had left was revenge.

Practice with his new body, that's what he needed, he decided. He took a hesitant step, then another. Walking was the first thing he needed to master.

He began to stride up and down the length of the laboratory. He would need coordination to escape. No matter how long it took, he would practice until he got it right.

The hours passed swiftly that night, but Flynn found marked improvement in his motor skills. Just before dawn, he returned to the corner where Grosswald had left him the night before. Let the doctor think he hadn't moved all night, and that he had no will of his own…revenge would come in time.

From then on, life fell into a simple pattern for him. During the day, he marched to Grosswald's drum, parading up and down, down and up, doing whatever the doctor said. Several times, high-ranking Nazis came to watch him. He always did his best to appear clumsy, once falling flat on his faceplate when ordered to run. He took a private satisfaction in Grosswald's embarrassment. Nevertheless, the Nazis seemed greatly impressed.

"It is a prototype," Grosswald kept reminding them. "Each new robot will be better than the last, until they are perfected!"

Night remained Flynn's own time. His coordination returned. He could walk, run, even skip if he chose, and though he knew he lumbered when he moved, it was swift and easy and purposeful.

Confident that he could act effectively, he decided to make his escape. But first he had to take care of Grosswald. The doctor could never be allowed to operate again.

To act effectively, he knew he had to end the doctor's control over him. The best way seemed to be to shut off his hearing. If he couldn't hear the doctor, he wouldn't have to obey: it seemed simple enough.

Raising his hands to his head, he poked a finger in each ear-slot, pressing down until the delicate receivers inside cracked and died.

Deaf, he returned to his corner and waited. Let Grosswald shout orders till he went blue in the face, Flynn thought with satisfaction. He wouldn't hear a thing.

Grosswald entered with his assistants just after dawn the next morning. As always, he crossed to Flynn, put his hands on his hips and shouted something.

Flynn didn't hear it, though. Instead, he took a step forward, reached down and grabbed Grosswald by the front of his lab coat. Lifting him as easily as a child lifts a doll, he carried him to the cell, tossed him through the open door and watched him slam against the bars on the back wall.

Then he turned toward the other assistants. They were gaping at him in shock. Taking half a dozen quick steps, he grabbed two of them, dragged them to the cell and threw them inside.

"Stay there!" he said. He couldn't hear himself, so he wasn't sure if the words came out. But when the two of them nodded quickly, the expression of panic on their faces was all the evidence he needed. They had heard him, all right.

The others were edging toward the door.

"Get in the cell," he said, pointing.

One by one they filed over to the cell, meek as whipped dogs. He counted seven, then frowned. Where was the eighth assistant? Probably sick or off today, he decided, scanning the room. Well, he had Grosswald; one assistant more or less wouldn't matter.

Closing the cell door, he gripped one of the bars in his metal hands, bent it out of position, then wrapped it around the door. That would hold them.

Using alcohol for fuel, he moved quickly about the laboratory, setting anything remotely combustible on fire. Flames spread up the paneled walls to the ceiling. Stacks of papers blazed in the corners. That should do it, he thought.

Opening the door, he strode out into the courtyard. It was a beautiful day. The sun shone brightly, a crisp new layer of snow covered everything, and the air had a perfect crystalline quality.

Turning, he looked back at the laboratory. A thick column of black smoke rose from the roof already.

Someone must have raised an alert; soldiers began to pour through the front gate and from buildings ringing the courtyard. Several carried rifles, and as they saw him, they began to shoot. He neither felt nor heard the bullets pinging off his body, but he saw sparks leap, and tiny dents and scratches appeared. Then a bullet hit him in the mouth, and his eyes flickered off for a second.

The sudden brush with blindness got him moving. He couldn't just stand here, he realized, and let them hammer away at him. He might be well-armored, but Grosswald couldn't protect every square inch of his body. They might accidentally hit something vital.

Turning, he ran toward the castle's open gate, his huge legs thudding on the ground. Everyone in his path turned and ran. Outside, he paused long enough to get his bearings.

On the other side of the forest at the foot of the hill, he could see rooftops and smoke from dozens of chimneys. A village—just what he needed, he thought.

He began to run again, his metal feet finding traction even on the snow-covered road. The villagers would help him, he thought. They had to.

He glanced back every few minutes, but found no sign of pursuit. The fire had bought him time.

At the foot of the hill, he left the road and pushed into the forest, heading straight for the village. Birds and small animals fled before him. He shoved trees out of his way, uprooting them. He'd known his strength was tremendous, but he had no idea it was this great.

Fifteen minutes later, he reached the end of the forest and emerged into the back yard of a large two-story tudor house. Children with wool hats and scarves were playing tag. They saw him, opened their mouths in screams he could not hear, then turned and bolted.

"Wait!" he tried to call. A stream of sparks shot from his mouth. The soldiers must have damaged whatever speaker allowed him to talk, he realized with dismay. How would he communicate with them now? Sign language, he decided—he'd have to act it out. That's what actors did, after all.

He followed the children around the building, out to

the cobbled street. Several dozen men and women in thick overcoats saw him and stared. They dropped their bundles and ran away in terror.

At the end of the street, the children had gathered around a policeman in a dark uniform with a spiked helmet on his head. He was shaking his head. Then the children began to point at Flynn, and when the policeman looked up and saw him, too, horror filled his face.

Flynn waved. The children darted off down the street. Seconds later, the policeman followed.

Flynn trailed them to the town square. A car idled to one side, its doors open. Several packages lay in the middle of the street as if hastily discarded. Not a living soul was in sight. Clearly they wanted no part of him. He would have to try to make it to the border on his own, he decided.

He turned and found a mob advancing on him, led by the policeman. There had to be thirty or forty men there, he thought, and perhaps more. Some were on foot, some in official vehicles. They were armed with pitch-forks, axes, clubs, even a few hunting rifles. As he hesitated, a bullet struck his chest and ricocheted.

Flynn didn't know what to do. He couldn't hurt them; they were innocent. No, he only had one option.

He began to run. His limbs might be untiring, and his strength might be huge, but he just wasn't built for speed, he realized as soon as he left the village. Even sticking to the road, making a beeline out of town, the mob was rapidly gaining on him. Each time he looked back, they were closer, fifty yards, forty, thirty—

He realized he wouldn't make it. From the crazed looks on their faces, there could be no reasoning with them, even if he'd been able to speak. He had to go someplace they wouldn't follow. A gravyard? None in sight. The forest? He'd leave a trail a blind Cub Scout could follow. Where?

He spotted a barn ten yards off the road. It was a ramshackle old structure, with peeling paint and doors that sagged on rusting old hinges. Would they follow him there? He didn't think so. And there might be a back door….

Leaving the road, he smashed through a wooden fence and crossed a field to the barn. The carriage doors opened easily; he slipped inside.

His gaze swept the hay-filled lofts, the empty stalls, the moldering leather tack hanging from hooks. No hiding places here, he thought. And no back door. He'd have to make his own.

As he advanced inside, something struck him from above. It was a piece of timber, he realized.

He looked up. A boy of perhaps fifteen or sixteen stood at the edge of the loft gazing down at him, a terrified look on his face. His shirt hung open and he had straw in his hair and clothing. Behind him, peeking out from the hay, was a blond, blue-eyed girl about the same age.

Young lovers, Flynn decided, on a romantic tryst… nothing to do with him.

Then he saw smoke and flames. The mob had found him. The villagers had set fire to the barn. They didn't know two kids were trapped inside. He hesitated. He couldn't leave them here. They'd die in the blaze.

Quickly, he motioned for the boy to climb down. In reply, the boy picked up another piece of wood and heaved it at Flynn, who batted it away with one hand. Well, they'd just have to do it the hard way, Flynn thought.

As flames climbed the barn's walls and began to kindle the hay above, Flynn crossed to the loft's ladder. More boards rained down, but he ignored them; mere wood couldn't hurt him, he told himself.

He tried to climb, but the rungs shattered under his feet.

His body weighed too much, he realized. Backing up, he motioned again for the two to climb down. Neither one did. Rubbing their eyes, coughing, they huddled together.

Flynn didn't know what to do. Thick, black smoke filled the air. He had to get them down or they'd be dead in minutes.

The hay they were lying on would have to break their falls, he thought. He smashed the stall partitions under them, then crossed to the oak beam supporting the loft. Slowly, carefully, he began to push.

Above, the loft began to creak. Hay rained down, some of it smoldering. Flynn continued to push, and suddenly, the beam snapped. The loft collapsed before him in an avalanche of hay and wood.

The two kids fell with it, landing on top of the hay, exactly as he'd planned. Neither one moved, but he thought they were unconscious from the smoke, not the fall. He had to get them out of here.

Picking up the girl—her body seemed so frail—he shielded her as best he could. Tucking down his head, he rammed the doors with his shoulder and burst through.

Villagers scattered like sheep. He set the girl down on the ground, turned, and ran back inside through a haze of smoke.

He found the boy on his feet, staggering. Flynn reached for him, but he jerked away, turned, and ran for the opening Flynn had made. Flynn followed, nodding to himself. This should buy him some sympathy from the villagers, he thought.

But he realized how futile it would be to go on. Tearing through the flaming back wall of the barn, hidden from the villagers by the inferno, he escaped into the woods. He was free—but the elation he felt suddenly collapsed into the worst depression he'd ever felt in his life. He was a machine now, and he would never be free again—unless….Slowly, resolve filled him. There was only one option. The Nazi's would never get their hands on Grosswald's creation, even if Flynn had to die to make sure of it. He searched his metal body for its power source. When he found it, he would rig it to explode. Flynn glanced one last time at the bright sun in the crystal sky, then ambled into the darkening forest.

Castle Grosswald
February 27, 1937

Heinrich Muller stood at attention before General Heimann. The general frowned as he surveyed the damage to Grosswald's laboratory.

"You're the only survivor," he commented.

"Yes, sir," Muller said.

"How is that?"

"I hid while the robot rounded up the other assistants and Dr. Grosswald. I was fortunate not to be discovered, sir."

"Ah." The general shook his head. "The doctor's notes were all destroyed in the fire?"

"Sir." Muller licked his lips. "I was Dr. Grosswald's second. I know every detail of his work. And I know what went wrong."

"What?" The general leaned forward, his curiosity evident.

"The brain...he left it free-thinking. It still had Flynn O'Conner's personality below the command overlay."

"This is bad?"

"Yes. He thought it would allow independent thought on the battlefield. I warned him it could lead to resentment and rebellion. He would not listen. He might have been a genius, but he refused to take advice."

Heimann nodded. "Very true. Go on."

"I know the project as well as Dr. Grosswald. I can continue where he left off."

"Very well," General Heimann said. "I will take up the matter with Berlin. In the meantime, find yourself a new laboratory. I want a new robot ready in two months for the Führer's personal inspection. And Muller...I will tolerate no more mistakes."

Muller nodded. "There won't be any," he promised.

War Department, Washington, D.C.
March 2, 1937

"Roll the film," Ian McBane said.

The room went dark, then the screen flickered and the newsreel began.

McBane sat silently as it played. It told of a real-life monster purported to have gone on a rampage through the sleepy German town of Brachtsburg only to vanish in a mine explosion that unleashed a terrible mushroom-shaped cloud. The voice-over made fun of the whole idea, as though the town had perpetrated a vast joke, but McBane knew better.

The footage ended and the lights came back on. McBane stood, cleared his throat and turned to his colleagues.

"My sources report some truth in this incident," he said, looking from face to face. "We also know our captured agent Flynn O'Conner was sent to Castle Grosswald several weeks before this alleged monster made his appearance. Castle Grosswald is two miles from Brachtsburg."

"What of O'Conner?" the White House liaison asked.

"One of our agents in the area recovered his body shortly after this alleged monster made its appearance, which is another reason we suspect Flynn of being somehow involved. That, and the fact that Baron Uwe Grosswald is dead as well. Heart attack, we're told. I don't believe it."

"Was O'Conner the monster?" the Secretary of Military Affairs asked.

"I'm not certain." McBane frowned. "An autopsy showed that O'Conner's brain had been surgically removed while he was alive. Dr. Grosswald was a renowned surgeon and roboticist. Is there a connection? I fear so. In fact, this whole affair smacks of a Nazi medical experiment gone wrong."

"What do you suggest?"

"I would like to develop more agents in this region. I want to keep a closer watch on Castle Grosswald."

"Agreed," the Secretary said. "I will see to the funding."

"Thank you."

As the others packed up their briefcases, McBane sighed inwardly. Flynn O'Conner had been a casualty of war, he told himself. The shooting might not have started, but the war certainly had. Hopefully, Flynn did not die in vain. Forewarned was forearmed, and suddenly, McBane wanted very much to know what had happened in Castle Grosswald.

Don't go near THE PANTANAL

Written by Karen Haber   **Illustrated by Rags Morales**

There were zombies in the badlands of Brazil, but reporter Shari Matthews didn't know that, yet. On the beach in Rio, all was sun-drenched torpor.

Suddenly, a dark cloud obliterated the sun.

What the hell? Shari wondered. January in Rio was supposed to be torrid. But all around her, beachgoers were shivering and complaining. A strange wind began blowing from the south, one that raised gooseflesh on her arms. It smelled of wetlands, of cold secrets, of danger.

A few drops of chill rain fell, spattering white sand, towels and blankets. But a moment later, the sun's golden rays broke through the shadows and the sky cleared, to cheers and applause.

Shari scanned the silvery waves breaking onto the white sand, leaned back in her beach chair and drained the last drop of her *caipirhina*, savoring the scent of the lime and the mule-kick of the *cachaca*.

Her watch said 1:20. She would have to be back at World News by two o'clock or her boss, Russ Albertson, would bitch.

That left at least half an hour for her to soak up the rays.

Behind her, Christ-the-Redeemer spread his concrete arms wide in benediction from his perch on Corcovado's peak, blessing the graceful skyscrapers of the South Zone district, the beachfront

traffic crawling, bumper-to-bumper, along the Avenida Atlantica, and the golden cariocas swaying sensuously down the mosaic sidewalk. Shari thought, not for the first time, that South America — and Rio de Janeiro in particular — was Heaven on Earth.

At ten after two, Shari was at her desk, cup of potent black coffee in hand. The robot dispenser had spilled at least two cups' worth on the floor before she had managed to get her mug under its spout. Dumb machine.

She scanned the latest wire service reports: Vietnam, Vietnam, Vietnam. Fighting was spreading to Laos and Cambodia despite the massacre of the Vietcong near Hanoi by U.S. forces. Nixon was spouting off — again — about the need for more, more, more funding.

Meanwhile, a tidal wave had killed 10,000 people in Bengal, women had just won the right to vote in Switzerland, and India and Pakistan were still duking it out. Same old, same old.

Shari cranked out local copy: A group of Guarani Indians on Brazil's southwestern border had vanished and ecologists blamed the Brazilian army. In Rio, police were gunning down beggars any time they wandered too far into the chic South Zone.

119

The phone cheeped twice.

"Matthews."

"Hello, luv." The clear lilting soprano belonged to Jax Hartley, attractive aide to the British cultural attaché and a disco-hopping pal of Shari's.

"Save me," Shari said. "I'm writing about the favelitas."

Jax sighed. "Too bleak. I'll trade you for the zombies."

"Zombies?" Shari said. "What the hell are you talking about?"

"No joke, sweetie. A friend just in from Sao Paolo told me there are zombies slogging all over the Pantanal. Creatures without souls, casually killing anybody they find."

Shari fed a fresh sheet of paper into her typewriter. "Zombies, you said? How many? And how many corpses?"

"Luv, I don't have all the details."

"Damn."

Jax laughed her familiar, musical trill. "Why not stop off down there on your way out of town?"

"News sure gets around." Shari felt a sudden chill. How had Jax found out about her upcoming trip to Uruguay? She hadn't told anybody. She and Kevin were just planning to slip away. Shari remembered the rumors that Jax was really with MI6 and her diplomatic role was just a cover. Why would Kevin confide in her?

"Listen," Jax said. "You could scoop everybody on this story. Or, you could just go lose some money in the fleshpits of Montevideo. Of course, if you get into any trouble you'll be on their own."

"Trouble? What kind of trouble can you get into playing craps?"

"Just a word to the wise. Darling, I must run. Call me when you get back to town. Kiss, kiss."

"Ciao."

Zombies in the Pantanal? Shari pondered Brazilian geography. The Pantanal was a stretch of hard country between Sao Paulo and the Uruguayan border, filled with alligators and poisonous snakes. She was tempted to cancel her trip with Kevin and fly down there to investigate this zombie rumor. It would make a great story. But it was probably too good to be true. The zombies would turn out to be a few drunken fishermen who

had wandered away from their party and frightened some superstitious villagers. No, no. She would go to Montevideo with Kevin, as planned.

Kevin Rogers was one of the crowd of American expatriates who were dodging the Vietnam War draft by partying — and gambling — in Rio. The city, in 1971, was an expatriate's delight, a wide-open, nonstop party where the police looked the other way as drugs and alcohol flowed freely. People with unsavory pasts could buy bright, shiny new identities in the shadow of the palm trees, and Americans dodging the U.S. legal authorities could spend their days on the beach and their nights at the discos.

Shari liked Kevin's blond good looks and sky-blue eyes. Liked, too, their plans for a weekend getaway of gambling and drinking. There was an aura of recklessness about Kevin, a whiff of danger that intrigued her. Oh, there was more to that boy than he let on, she was sure of it. Yes, she would definitely go to Uruguay.

The Varig flight took two hours. As they passed over the Pantanal, Shari stared down at the rocky, emerald wilderness and imagined zombies tangling with giant alligators in the green darkness below.

They were soon on the ground at Montevideo International Airport and into a taxi.

"*Buenos días,*" said the driver.

"*Vamanos a Rambla Parque Hotel,*" Kevin said.

Shari found it strange to hear the sharp-edged tones of Spanish after her immersion in the soft slurring of Brazilian Portuguese.

Montevideo was a graceful well-kept city of parks and plazas. Noble statues of obscure politicians rose up everywhere. The Rambla Parque Hotel was a handsome pink beachfront tower with the Aquarius Casino next door.

The casino was lush and cool, a place of sparkling lights, soft pop music and the sounds of people at serious play. The gaming tables were green felt, edged by white leather embossed in gold with the signs of the zodiac. The chairs were well-padded, and the robowaiters could not have been more obsequious. Shari had only to look up from the blackjack table, and a robot with a tray was at her elbow, offering her another *Cuba Libre*. She finished her first rum-and-coke in four gulps, irritated. Kevin's best buddy, Steve Sherwood, had horned in on the trip, showing up unexpectedly.

Shari ordered another drink and adjusted her white hip-huggers. The mirrors on her sleeveless Indian vest winked, pink and blue, reflecting the casino's spotlights.

Kevin was intent on the game. He had started out hot with a winning streak, and after each successful hand, he pulled her close for a kiss. "For luck," he said. Shari played a round or two but got restless. She had plans for them upstairs, behind closed doors.

Kevin pushed back from the table. "I'm going to make some phone calls," he said. "There are some guys trying to duck me. They owe me a fortune."

Shari shrugged away her disappointment. She knew that he got by on his winnings — and other dealings. "Make it quick, okay?" She brushed his lips with hers.

He gave her a wink. "Only a crazy man would stay away from you a moment longer than he had to." He squeezed her briefly and sauntered away.

I'm not in love, she told herself. Just in lust.

A tall, heavy-set man with white hair leaned in Kevin's direction. They exchanged a few words. Kevin shook his head. The man grabbed his arm. Kevin yanked himself free and walked away. The stranger glared after him.

What the hell? Shari wondered.

"Been dumped?" Steve Sherwood, dark, bearded and serious, fresh from baccarat, smiled down at her. Shari grinned. "Yeah. Kevin's looking for some pals who owe him money."

Steve rolled his eyes. "I've heard that one before. Well, how about keeping me company until he gets back?"

"Best offer I've had all night."

He frowned. "Hard to believe."

They dabbled with roulette, moved on to poker until Steve's luck, hot at first, gradually began to cool. Finally, they came to ground back at the blackjack table. Shari won a nice sum but Steve lost several hundred dollars.

He shook his head ruefully. "Gambling's a tough way to make a living."

"Kevin says it's better than slogging through the mud looking for strangers to kill."

Shari glanced at the clock. Kevin had been gone for over two hours. "Where the hell is he?"

"Speak of the devil."

It was Kevin, shouldering through the crowd of gamblers and heading in their direction.

Shari planted herself squarely in his path. "Hello, stranger."

The look in his eyes — flat, dead, blind — chilled her to the core.

"Kevin?" He moved past her and disappeared into the bar.

Stunned, she stared after him. "Hey! What gives?"

"Something's up," Steve said. "I'll talk to him."

Shari throttled her confusion and irritation by shooting a few rounds of craps and winning a few more dollars. But that soon paled. She decided to go see what was up herself.

As she entered the dim, smoky room, brilliant fireworks — after-images from the casino's pin spots—exploded against her eye-lids, blinding her for a moment. When she could see again, she scanned the tables and booths. No sign of Kevin or Steve.

What in hell was going on?

"May I buy you a drink, señorita?" The speaker was a short man in a white nehru jacket with a florid complexion and long, greasy-looking hair.

"No, thanks." Baffled, angry, she strode out of the bar. What now? She went upstairs. Maybe he had returned to the room. No. The room, smelling faintly of perfumed disinfectant, was empty.

"To hell with him!" In anger and disgust she tore off her clothing and collapsed into bed.

She slept restlessly and awoke early, still mystified. How could Kevin and Steve have vanished like that? This was no simple kiss-off. She phoned Steve but there was no answer. He didn't respond to a knock on the door, either.

A petite maid with apple cheeks came around the corner toting a vacuum. "*Buenos días.*"

Inspiration bloomed. If Shari could convince the maid that she had spent the night in Steve's room and left her keys there, perhaps she would let her in to look around.

A nod, a wink and the quick transfer of a few coins. It was understood, a matter between women. The door snapped closed behind her. Shari turned, flicked on the light and gasped.

The place was a horrifying mess. The furniture had been smashed to toothpicks, the bedding and drapes shredded as though by some relentless monster. Who — or what — could have done this? There were no signs of blood. But obviously somebody — Steve? — had put up a fight. On the floor, by the remains of a lamp, was a crumpled piece of paper. On it, a scrawled address, a place somewhere along the waterfront, Shari suspected. Worth a look, she thought.

She went out. The maid was still in the hallway, running the vaccum over the carpets. Had she heard anything or seen anything unusual? No. *Gracias.*

The address on the paper was a long taxi ride away, taking Shari through the neat streets along the beachfront to a dusty warehouse district near a series of docks.

The driver asked her if she was certain this was the right place.

His frozen expression never changed. She tried to pull her arm free. His grip was like steel.

"Let me go."

There was no answer.

A rag with a sickly-sweet substance was clamped over her face. She couldn't breathe. Then everything went dark.

The room swam slowly into focus. Shari was lying on a hard metal surface, and her head throbbed mercilessly. There was a peculiar institutional smell that she associated with girls' bathrooms in high school.

In the dimness, she could make out strange objects, electronic equipment, switches, dials. Some sort of laboratory, apparently.

Was this where Kevin had been stripped of his free will? Was she going to be zombified as well? She struggled to sit up. Her arms and legs were bound with rope.

Thump!

What was that? To her left — strange, insistent thrashings.

"Psst! *Señorita. Por favor! Habla usted espanol?*" The voice was male, high-pitched and quavering but urgent. "*Habla espanol?*"

"*Un poco. Habla ingles?*"

"Yes, I learned it in school."

Shari lifted her head until her neck hurt. She could just make out a dark form lying upon a table across the room, trussed like a bird for the slaughter.

"Can you move at all?" she asked. "Toward me?"

"I think I can swing the table a little. It's on rollers." It took a maddeningly long time for him to get moving. He leaned hard to the left. The table gave a bit. He swung himself in the other direction. The table rolled a bit more. He wiggled, nearly flipping onto his side, and succeeded in moving the table a few inches closer to Shari.

The wheels squeaked like angry rats. It seemed all too likely that the noise would attract the attention of their jailers. Frantically, Shari began to rock to and fro, trying to move her own perch toward him. The table rolled a little bit. Heartened by that, she rocked with more intensity. Rocked and rolled.

The tables met with a crash in the middle of the room.

A wiry, young Latino man of perhaps eighteen stared at her from inches away with frightened eyes.

"I can try to pull your ropes with my teeth," Shari said. "Can you move any closer?

She managed to get up on her knees and hunch over him. It was an embarrassingly intimate position. Her hands were beginning to go numb, but she forced herself to ignore that and worked on her fellow prisoners' cords, biting and ripping at them until her jaws ached. The shredded fibers tickled her nose unmercifully, and she felt as if she was about to sneeze explosively.

She nodded and waved him on. The cab disappeared in a cloud of dust.

Her platform heels made a loud and lonely clip-clop on the cracked pavement. The warehouse doors were locked, the windows shut and barred. She turned a corner and found herself in a dim, garbage-strewn alley.

Halfway down it she realized that she heard a second set of footsteps echoing her own, then a third. She whirled, but hands were already grabbing her.

"Hey! What the hell —?"

She struggled frantically, screaming for help. She twisted around somehow and confronted her attackers. Then her voice died in her throat. "Kevin?"

Kevin, yes. He was gripping her left arm, looking strange, robotic, almost inhuman. An unknown man with black hair was holding her right arm.

"Kevin?" It was a whisper that she could barely force out. "What's happened to you?"

Then, at last, one of the cords broke.

"Keep going!" whispered her fellow prisoner.

She broke through another cord and another. He began to wiggle his hands frantically. In another moment, they were free. Quickly, he leaned over and untied Shari's hands before moving on to the bonds on his legs.

"My name is Andres," he said. "I'm with the ecology movement in Argentina. I came to Montevideo for a conference. But my friends disappeared after the first night. They've been missing for two days, and I've been looking for them everywhere."

"I don't understand. Then why are you here?"

"I was attacked by terrible men with dead eyes. They grabbed me. I was fingered by this *puerco* of a German industrialist, who hates all of us trying to shut down his polluting factories. I woke up here."

"I'm Shari." She rubbed her wrists, trying to get circulation moving. "I work for the World News Network. I was down here with my boyfriend. Approximately the same thing happened to me."

There was a sudden noise, regular, getting louder. "Footsteps," Shari said. "Shhh! Hide!"

They slithered off the tables and under a pile of braided cords and padded mats.

A door creaked open. Lights flickered on in the front of the lab. But their hiding places were cast in shadow. Two men in white lab coats entered the room, towing a gurney behind them. There appeared to be a body on it, draped in a grimy cloth.

The men seemed oblivious to what was around them, intent solely upon a cluttered worktable. They spoke quietly, in English, of sorts. One of them, a stout grey-haired man, very old, with a small, clipped mustache, had a heavy German accent.

"I told you not to use that circuitry," the German said. "It feeds back too much power and fries the unit."

His colleague muttered apologetically.

"And what of the report on our experiments? Have the cyborgs returned from the Pantanal, yet?"

"All but three."

"Three? What happened to them?"

"There's no sign. We've flown reconnaisance on them but we can't pick up their heat signatures."

"Idiot! This sort of sloppiness will not be tolerated!" The German's voice crackled. "I was taught discipline by Dr. Grosswald, and I will have it here in my lab. Do you understand me?"

The smaller man shook visibly. "Yes, Dr. Muller."

"We're close to success, very close. There must be no foul-ups. If three renegade cyborgs have somehow gotten loose in the Pantanal, they must be found and destroyed. Understand? We can always create more. We'll be moving to Matto Grosso in a few months. I want them perfected by then."

His words had a deadly ring. Cyborgs. Didn't that mean human robots? Zombies?

They're making their own army, Shari thought. And making it out of people that nobody looks for: Indians, draft-dodgers, ecology freaks — throw-aways and trouble-makers. Expendable people.

Dr. Muller appeared to be adjusting a small square device, some sort of handset. He aimed it at the prone figure on the nearby table. The body raised an arm, then a leg.

Muller nodded. "Good. Grosswald never dreamed of the possibilities inherent in controlling the actual human body and bending the will. If only he could be here to see this."

He conferred with his assistant in murmurs too faint to hear. Then he put down the handset. "Enough."

The two men grabbed the gurney upon which their subject lay and pulled it out of the room, slamming the door behind them.

Shari sprang forward the moment they were gone. She seized the handset, puzzling over it, trying without success to make sense of the cryptic symbols that marked its surface. Then, abruptly, she slipped it into her pocket.

"What are you doing?" Andres whispered. "What do you want that for?"

"It might be useful somehow." She peered across the room. "Let's try that door."

It was locked. "There has to be some way out of here."

A supply cabinet had a faint outline of a door cut into its shelving. Shari pushed against it, and the wall swung open onto a dim corridor. "Come on!"

They groped their way down the passage, moving toward a lighter hallway. This time they found a door that wasn't locked. It opened onto a deserted and dusty alley at the back of the lab. Shari heard a faint shrilling sound, growing louder. An alarm bell. And behind it, a regular beating sound. Footsteps.

"They're on to us," she said. "Move, move, move!"

They careered around the corner into a wider street. An abandoned newsstand leaned drunkenly against the far curb.

Shari glanced back over her shoulder. Dark shadows came around the corner and spread into the street. She could see a group of men moving awkwardly, lumbering toward them, five, six, eight of them, maybe. Were they cyborgs? She could make out the features of the ones in front, now. Oh, God, she thought, it's Kevin. Kevin and Steve.

"It's no good," Andres said, clutching Shari's arm. "There are too many of them."

Shari ducked into a doorway behind the newsstand but Andres bolted down the street. The cyborgs thundered after him, emitting an eerie droning sound.

One stayed behind. Kevin. He lunged at Shari.

She put the newsstand between them. He came around the side.

Desperately, she tried to break through his trance. "Kevin! Kevin, listen to me. It's Shari. Don't you recognize me? You know who I am!"

His robotic gaze never flickered.

"You don't want to hurt me," she said. "Look at me! I'm Shari!"

Kevin reached for her, grabbing at her hands. She pulled back out of reach. He continued toward her. Recklessly, she slapped him, hard, across the face. "Dammit, Kevin!"

He made no response, not even a frown. That was the most terrible thing of all. But the neck of his shirt had fallen away and there, glinting blue-black on pink flesh, was an awful device, a metal spider whose cables disappeared under the skin. A control box.

Shari tore her eyes from it. She made a break for the broad street in front of the lab. Kevin grabbed the back of her shirt. It began to rip. His other hand closed on her shoulder.

"Kevin," she cried. "Remember all of our good times in Rio? Gambling? The volleyball games?" She knew that she was babbling. It wasn't working. What about the device in her pocket? Would that control him? She had to try it.

Twisting in Kevin's grasp, Shari yanked the handset out of her pocket and jammed it against his chin. It was covered with rows of blue and red buttons. She began to push them at random.

There was a faint whining sound.

Kevin let go of her and lurched backward.

She kept pressing buttons.

Kevin jerked and twitched like a marionette. Shari could hear the other cyborgs returning. She spun around to train the handset on them.

They fell into chaos, puppets whirling left and right, crashing into one another. Shari pressed the last row of buttons. A moment later, the handset went flying, knocked out of her grasp by Steve. It fell to the ground. His foot hovered over it; then he brought it down on the gadget with crushing force.

Shari didn't stick around to watch. She dashed down the street, heart pounding. In a moment, Kevin or Steve would be upon her.

But Kevin wasn't moving. He stood at the mouth of the street and shuddered. He raised his hands to his face and moaned. His eyes flickered, coming to rest on her face. "Shari?" he said dazedly. "Where are we? What's going on?"

"Kevin, Kevin, thank God." Shari fought back tears of relief. "You've been hypnotized or something. Steve, too. We're trying to help you, to escape."

"Save me? Escape? From what?"

"It's a long story..."

Steve smashed into Kevin, knocking him to his knees. "Hey! What the hell?" Kevin was staring a nightmare in the face. "Steve?"

Steve took aim at his jaw. Kevin managed to parry the blow and used Steve's momentum against him, yanking his foot up and back. Windmilling his arms, Steve fell hard.

"What's wrong with you?" Kevin demanded.

"Don't ask questions," Shari cried. "Just run!"

Kevin, still half-dazed, was clumsy and slow.

Shari had hoped to flag down a passing motorist, but the road was empty. She cast around. There, leaning against the curb, an old, rusted Volkswagen, with Andres hiding behind it. "Andres," she cried. "Help me get into that."

Andres nodded. He picked up a large stone and began pounding on the passenger door window until it shattered. In a moment, he had it open and had unlocked the driver's door. Shari crawled in.

"I think I can pop-start this crate if we can roll it."

Andres leaned hard against the doorframe, attempting to push the car. "It's too heavy."

A sudden jolt gave the car a brief spurt of momentum. Kevin stood behind them.

"Get out of here," he said. "Forget about me."

The cyborgs came around the corner, fully recovered.

Shari glared at Kevin. "What are you talking about? Get in the car, now!"

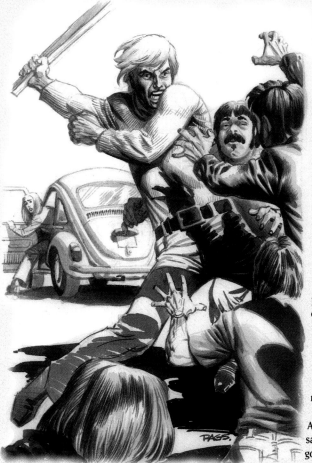

"No, they'll catch us." His words were slurred. He was still half-zombified, barely functional. "Get away and spread the word. This goes much deeper than you think."

Andres leaped back into the car. "*Vamanos*," he said angrily. "We don't have time to waste."

"Kevin, please!"

"Somebody has to push the car," Kevin said. "Get out of here." He slammed her door shut and gave the car another shove. It began to move. He shoved it again.

The zombies were closing on Kevin, with Steve out in front. Shari put the car in first gear, stomped hard on the clutch and prayed.

The engine snorted. Steve reached for Kevin's arm.

Kevin shook him off and pushed the car again.

It scooted forward, picked up speed and began rolling down a slight incline. Shari fed it some gas, then stepped carefully upon the brake, lifting her heel almost immediately.

The engine turned over.

She pumped the gas pedal. The engine coughed, turned over again, caught.

"It worked," she cried. "Kevin, run and grab hold! Andres will pull you in on his side."

But Kevin wasn't listening. He stood in the dusty street, gave a half-wave and turned to deliberately face the mob.

Shari watched him disappear under a ferocious wave of zombie cyborgs. Tears flooded down her cheeks. Half-blind, she drove like a madwoman, taking corners on two wheels, gunning the rickety car through the unfamiliar streets.

For a time, neither she nor Andres could speak. Finally, he said, huskily, "Where are we going?"

"To the nearest police station."

"No! The Uruguayan police won't listen. They're as likely to throw us into jail as listen to a wild tale about zombies on the docks of Montevideo."

"Then where should we go?"

"The airport."

"But what about my friends? And yours? We've got to save them."

Andres stared at her. "Are you crazy? Go back there?"

"With the police."

"They won't come." Andres' tone was caustic.

"You North Americans don't understand what it's like down here."

"But Kevin..." She couldn't finish.

He patted her arm. "I know. But he's dead or as good as. My own friends, too. The best thing we can do is get away and try to spread the word."

The next morning, bleary-eyed, Shari marched into Russ Albertson's office and slammed the entire story down on his desk. She waited as her boss read it, then said, "Well, what do you think?"

Albertson squinted at her over his unlit cigar. "I think it's a load of bullshit. What do you expect me to say?"

"Hey, wait a minute —"

"I guess it's okay as science fiction. But we don't publish science fiction here."

"Russ, it happened."

"Nobody will believe it. You can't seriously expect me to run this crap. Next you'll be telling me you saw former S.S. members on the beach at Copacabana."

"I thought that this was a news organization."

"It is. That's why I'm doing you the favor of destroying this tripe."

As she watched in horror, he tore her story in half and dropped it into the wastebasket.

Stunned, Shari turned and walked back to her desk.

She thought for a moment, nodded and began to dial.

"British cultural attache's office. Jax Hartley speaking."

"Jax, it's Shari. We need to talk."

"Darling! Why not meet me at the Rhinoceros tonight, at nine? And bring that handsome devil Kevin along."

"Kevin is dead."

"What? Where? How?"

"It's a long story and a crazy one. Remember those zombies you mentioned, Jax? In the Pantanal? Well, they're real. They're real, Jax, and they're in Uruguay, too. They killed Kevin."

"Shari, did somebody slip mescaline into your *Cuba Libre*?"

"Dammit, this is serious! I don't know who else to go to. I don't know if I can even trust my own government. Jax, you're it."

There was a pause. When Jax spoke again there was no trace of gaiety in her voice. "My office. Fifteen minutes."

"Good. I'll be right over." Shari hung up and told herself that the counterattack was just beginning. I'll go to the Embassy. The CIA. I'll move Heaven and Earth. And I'll find out how Kevin fit into all of this.

She gazed out the window, scanned the carefree beachgoers down below as they played in the silvery waves, oblivious to their risk, and thought, for the first time—but not the last—that South America was Hell.

JOURNAL

JOURNAL ENTRY

Entry 1, 12/14

    In order to build things up, often we must first tear them down. This is painfully clear to me, now that everything I thought I knew about grandfather and my family's past has been completely upturned. Now, that darkest evil has pursued me across half a world.

Now, that Jean is gone.

    Still, I consider myself fortunate that through the cloud of sadness which envelopes me there remains a beacon of hope — in confronting the truth, I have achieved a new kind of freedom. The freedom, I hope, to make this world better, to make it a world where a devilish machine like the Iron Major could not survive one day, let alone fifty years. A world where a woman like Jean would not feel compelled to dress herself in deception and espionage.

    It's a comfortable feeling...as if I have taken up the work my grandfather put down so many years ago. Work, perhaps, I was meant to do. And, if I can learn, as well, from the savage melding of human life and mechanical form that drove the Man of Iron and use the knowledge beneficially... then that will be the ultimate retribution against the brutality of the Nazis. And some small measure of vengeance for Jean.

    I have learned the importance of historical perspective, of understanding the chain of actions and events in the past that makes us who we are today. Now, my task is to follow this chain to its brightest possible conclusion. Towards that goal I have started this journal to record the people and places that have gone before me and to remind me of the dangers of misguided science. My work seems more important to me now, than ever before.

    Science was innocent once. Jean said that to me, and for her, I will try to reclaim that innocence.

TIMELINE

1922 First successful independent robot produced.

1927 Publication of the Lang Protocols by Pro. Eando Lang — three tenets designed to govern the behavior of robots for the protection of humans. Though widely regarded, these guidelines were often ignored because of their inherent commercial limitations in such vast markets as the military and law enforcement.

1929 First commercial use of crude independent robots. These were available only to the very wealthy or to major corporations and served mainly as novelties.

1932 Colin Brennan founds Interworld Technologies, which eventually becomes the leading robotics and technologies corporation in the world.

1935 German robotics program launched and headed by Dr. Uwe Grosswald with a mandate from Adolf Hitler.

1937 German robotics program creates an advanced, independent robot prototype capable of accepting crude organic input, but the father of this machine, Dr. Uwe Grosswald, disappears under mysterious circumstances.

1939 World War II begins.
New York World's Fair presents "The World of Tomorrow," which features among its many attractions a remarkable, new robot prototype sponsored by the General Energy corporation. The robot later vanished and was never recovered.

1945 World War II ends.
Unsubstantiated reports indicate possible robotics experiments conducted by Nazi researchers in Switzerland as a last attempt to turn the tide of the war.

1950 Korean War begins.

1952 First effective use of robots in military action during the Korean War with minimal deployment due to high production costs.

1954 United States Robotics Authority established by robotics researchers and corporations in reaction to the "Steel Scare" as US citizens feel increasingly threatened by smarter and better robots. The USRA became responsible for maintaining the Lang Protocols within the robotics industry.

1956 Congress passes legislation restricting use of armed, non-protocol robots to the military and law enforcement authorities.

1964 An incident in the Gulf of Tonkin sparks United States involvement in the Vietnam War. Robots are deployed in combat immediately with devastating results.

1971 Unsupported rumors of zombies in South America coincide oddly with reported sightings of Heinrich Muller, a Nazi believed to have survived World War II and escaped prosecution for war crimes. Muller was Uwe Grosswald's assistant before the war.

1973 The United States and South Vietnam win a decisive victory over North Vietnam.

1976 Patrick Brennan elected President of the United States.

1989 Annabelle Brennan assumes control of Interworld Technologies and changes the name to WorldTech International.

1992 Michelle Timmons elected President of the United States.

1995 Professor Zac Robillard creates the world's first I·BOTS, truly independent robots capable of thinking for themselves. I·BOTS are completely bound by the Lang Protocols.

THE GOLDEN ROBOT / GOLDEN BOY

Date of construction: 1939
Modified: 1945
Robotics Designer: Unknown — Modified by Aaron Robillard and James Creed
Point of Origin: USA

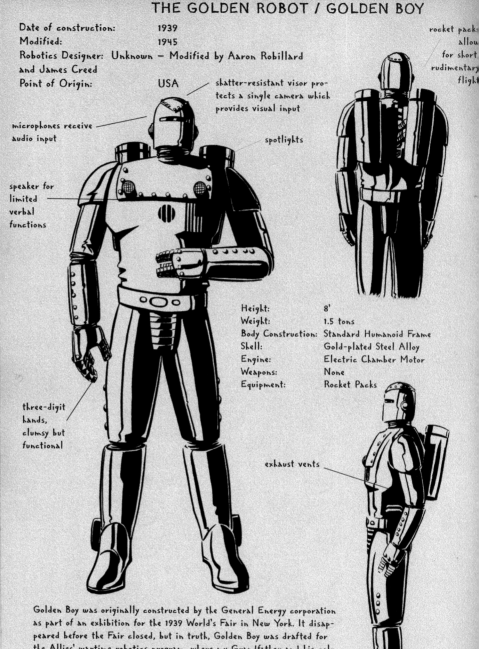

rocket packs allow for short rudimentary flight

shatter-resistant visor protects a single camera which provides visual input

microphones receive audio input

spotlights

speaker for limited verbal functions

three-digit hands, clumsy but functional

Height: 8'
Weight: 1.5 tons
Body Construction: Standard Humanoid Frame
Shell: Gold-plated Steel Alloy
Engine: Electric Chamber Motor
Weapons: None
Equipment: Rocket Packs

exhaust vents

Golden Boy was originally constructed by the General Energy corporation as part of an exhibition for the 1939 World's Fair in New York. It disappeared before the Fair closed, but in truth, Golden Boy was drafted for the Allies' wartime robotics program, where my Grandfather and his colleague, Prof. James Creed modified it for combat. After recovering the robot, I have made extensive repairs and improvements, but the very fact that it is functional at all after nearly fifty years is a testament to the genius of my Grandfather and Prof. Creed.

fins allow robot to steer during flight

THE NAZI ROBOT / THE IRON MAJOR

Date of construction: Unknown
First Activated: Unknown
Robotics Designer: Dr. Uwe Grosswald
and Dr. Heinrich Muller
Point of Origin: Germany
and Switzerland

advanced rocket
allows for pro-
longed flight

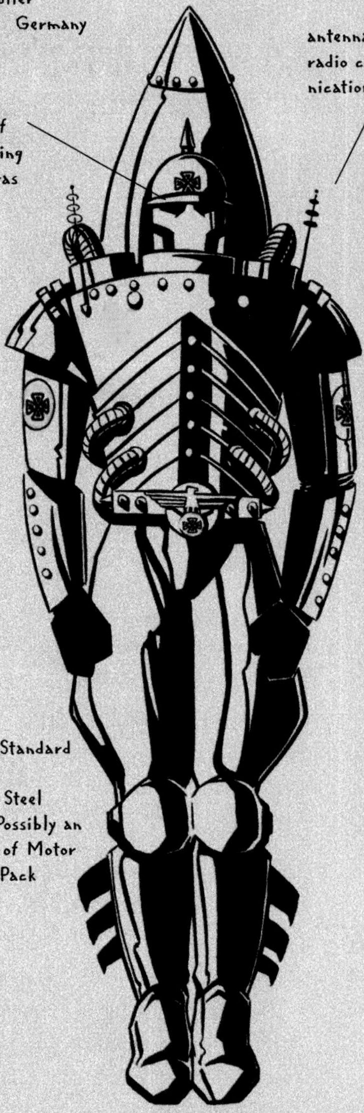

antennae for
radio commu-
nications

bullet-proof
glass covering
twin cameras

Human brain housed and
protected inside head

Height: 8'
Weight: 1.9 tons
Body Construction: Standard
Humanoid Frame
Shell: Matted Grey Steel
Engine: Unknown; Possibly an
Experimental Form of Motor
Equipment: Rocket Pack

exhaust tubes
circulate hot
air through
cooling system

directional fins

The Man of Iron was a prototype for what the Nazis hoped would become
an indomitable force of mechanical soldiers, which would turn the tide of
the war. They failed in that regard, but the Man of Iron has been active
covertly since the end of the war. Notable from a scientific view is the
fact that this is the only known robot to successfully interface a robotic
system with a human brain as it's central processor. If it wasn't a thing of
pure evil, it'd be a marvel. Still, the science behind it may prove important.

Professor Creed in the lab
Revere, MA, early 1939

My grandfather at work
Hembley Manor - 1945

Hagen --?
Date unknown

Jean

Keating Lab,
Demonstration of
Basic Organic Coding
Process #1118

Skinburn College, 1992

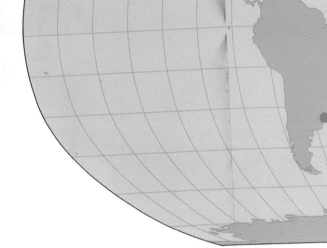

GERMANY

Berlin — Center of Nazi activity during World War II, primary staging ground for wartime robotics research... though major work was conducted in top secret locations around the country

Brachtsburg — Home of Castle Grosswald — used by notorious Dr. Uwe Grosswald as a robotics lab before the opening of the war. Records show Grosswald's breakthrough work led directly to the technology used in the Iron Major, but the lab closed mysteriously in 1937 — possible connection to the project that built Iron Major...?

Brachtsburg is now a minor tourist attraction based on local legends of a strange, semi-human creature stalking the nearby mountains.

SWITZERLAND

Argau — Grandfather's notes identify a chalet in this province as the site of the Golden Robot's first battle against Iron Major — both robots vanished afterward.

The Golden Robot resurfaced in a cemetery outs____ village in Argau — apparently hidden for decade____ in the grave of a war hero...

ENGLAND

Skinburn College — The ideal place to pursue m____ in robotics and build upon the work I've already____ molecular biology... and, of course, what good is____ without an understanding of history, ethics and ____ to guide it.

Hembley Manor — MI6 robotics lab hidden in ____ manor — here, I rebuilt the Golden Robot. Sho____ been secret, but word slipped out. MI6 is a go____ agency, after all...

UNITED STATES

New York City — The Big Apple! Site of the ____ Fair — "The World of Tomorrow." Hugely popu____ Energy exhibit featured "Mechano" — design ____ facture form a striking precedent to our US ____ History: Nazi Fifth Columnists stole the mac____

was destroyed during a recovery attempt.
Intuition: "Mechano" fell into the hands of the
Federal government...
Detroit — Danbury Mechanical Man, 1922 —
first independent robot created in a Danbury
Motors, Inc. research lab. Possible connections?

● URUGUAY
Uruguay — stronghold of the Iron Major. Many
Nazis fled here after the war and the Major
organized them toward his goals — they built a
fortress deep in the rain forest. Records recovered from the ruins show construction began in
mid-60s and ended in 1973.
Pantanal/Montevideo — M16 recorded numerous
zombie sightings in this region during the years
when the Fortress was being built. Mere superstition, I'm sure, except for one noteworthy
account from a World News reporter — claimed
Nazi scientists conducted devilish experiments
here in 1971 — insists they brainwashed many
people, including her boyfriend, changed them
into mindless cyborgs under their control. Most
probably pure fantasy... but if so, then why has
this case never been closed...?

● VIETNAM
Rumors still persist that "cybervore" technology
gave the US it's decisive victory here. No scientific records substantiate this. Where is this
technology? Could it be connected to Muller and
the Nazi's robotics experiments? Must investigate... Fen Burleigh... Exterminating Angel...?

○ KOREA
1952 — first effective use of robotics in warfare, primarily robotic tanks and gun installa-
tions deployed on a limited basis. Proved
effective... but cost of production prohibited
mass implementation. This and their relatively
late entry into the conflict kept them from
turning the tide of the battle... connection to
Muller or the Major?

Illustrated Novels:
A New Art Form for a New Age

They say a picture's worth a thousand words...and they're right. As anyone who's seen a movie and then read the book it's based on can tell you, every media has its own strengths, things it can do better than any other form. And that's what Illustrated Novels bring: the strengths of two different media—strong, elegant, cinematic prose and detailed illustrations that speak volumes to the reader.

The power of prose combined with the beauty of rich illustrations.

ISAAC ASIMOV'S I•BOTS

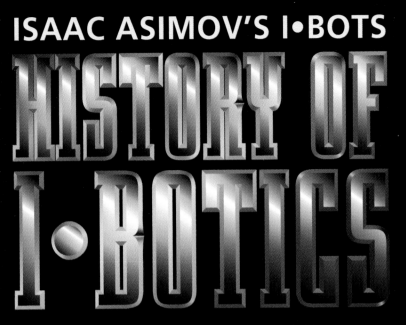

HISTORY OF I•BOTICS

⊸ᴇᴡᴏ⊸ AN ILLUSTRATED NOVEL ⊸ᴏᴡᴇ⊸

HarperPrism

A Division of HarperCollinsPublishers

Edited by James Chambers
Designed by Cynthia de Vosjoli & Tom Koziol

HarperPrism

A Division of HarperCollins*Publishers*
10 East 53rd Street, New York, N.Y., 10022

HarperPrism books may be purchased for educational, business, or sales promotional use. For information, please write: Special Markets Department, HarperCollins Publishers, 10 East 53rd Street, New York, NY 10022-5299.

ISBN: 0-06-105539-5

Printed in the United States of America
First Printing: August 1997

97 98 99 00 01 ❖/RRD 10 9 8 7 6 5 4 3 2 1

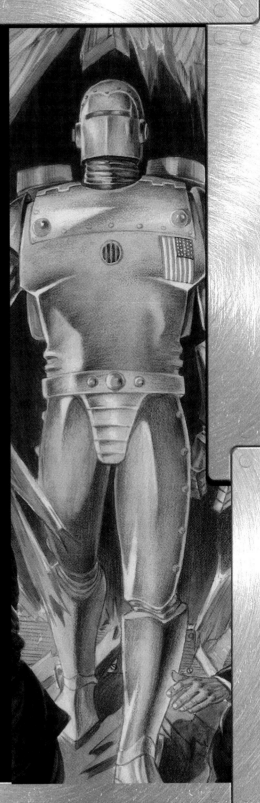

ISAAC ASIMOV'S I•BOTS
HISTORY OF I•BOTICS

❀—❀

❀—❀

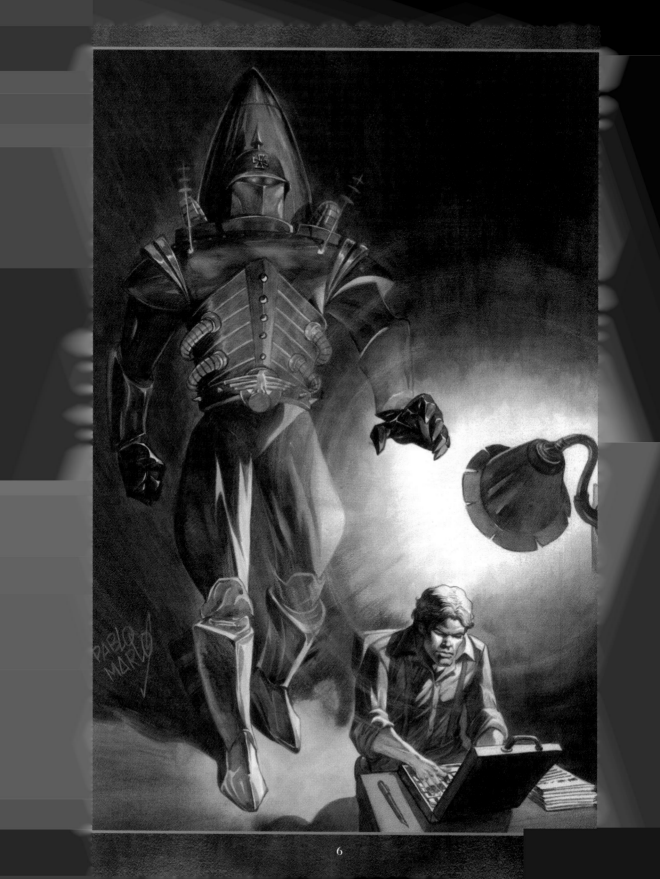

A NOTE TO THE READER:

When Mitchell Rubenstein and I were founding the Sci-Fi Channel in the early 1990s, we needed some advice, and decided to establish a Board of Advisors to link our efforts to the science fiction writing and fan communities. We already had Martin H. Greenberg, SF expert and the world's most prolific anthologist, at our disposal, and Marty mentioned that he had a friend who knew everything about science fiction and was world-famous to boot—Isaac Asimov. Isaac not only joined our board, but he brought his friend, the legendary Gene Roddenberry, along.

All of us loved Isaac, and when the Sci-Fi Channel went on the air in September 1992, the premiere was dedicated to the memory of Isaac and Gene, both of whom tragically died prior to the channel's launch.

But both of these great men had given us even more—ideas and vision that we were able to translate into great storytelling for our new venture, BIG Entertainment. In the case of Isaac, we had asked him to give us some ideas for a science fiction "hero" who could be taken into young adult and illustrated novel formats; and true to Isaac's genius, he came up with exciting fresh ideas that formed the basis of the "I•BOTS universe."

Isaac has always been fascinated by robots. When he was nineteen, he witnessed a clunky, mechanical man on display at the 1939 World's Fair, and much of his own work involved his wonderful "Positronic" robots and his seminal *Three Laws of Robotics*. *I•BOTS* however, is different. This was Isaac's chance to play with some other concepts—and while nothing in *I•BOTS* contradicts his *Three Laws*, this story gave him a chance to take things in a slightly different direction.

We've had great fun playing in this universe he created. We believe that he would be delighted with what we've done with it, and we hope that you will be too. And, just like the Sci-Fi Channel, this book is dedicated to the inspiring and beloved memory of Isaac Asimov.

Laurie Silvers, President, BIG Entertainment

Division

MEMO

To: T. Severn
From: Classified

Re: Lost Boys
Date: Classified

I've reviewed the matters you've brought to my attention. I was quite surprised to see these cases cross my
desk, again. Honestly, at this late date, I'm not sure why
you expect to find anything new in them. I still feel that
the incidents are connected, but the exact nature of the
connection defies analysis with the information at hand.
    But, since you asked...
    Whatever Grosswald was working on before World War II
certainly had repercussions well beyond the war years and
into the present. Unfortunately, we know little of the actual
events that stemmed from his experiments. As you know, the
Nazis opted for a low profile after the war. Your boys will
have to dig a bit deeper to shed any new light on the
activities of their scientists. Have you checked with the
Americans on this? Project Paperclip might've been an
attractive cover for some of the men you're after, and if
you approach Benson the right way, he just might agree to
help you.
    Otherwise, I suggest (as I have several times already)
that you pick up the trail in Uruguay. Regardless of how
fantastic the reports sound, it's entirely possible that
Muller relocated there to carry on Grosswald's work.
Besides, it may be the only lead you have. I still don't
understand why you've let it grow cold for so many years,
especially after what was lost there. Trust me. If you
ignore it, it will come back to haunt you.
    I understand this one is close to your heart because
of your connection to Robillard and Creed, but you've got
to be willing to take some risks if you want to get anywhere. If you're convinced that someone is actively continuing Grosswald's work, why not try the obvious. Use the
grandson to draw him into the open and stir things up a
bit. I know it's a distasteful prospect, but you certainly
have the stomach for it.
    Well, that's all I have to say on this. And I mean
that literally. I don't want to see these files again,
unless they're brimming with new intelligence.
    Good luck, Severn.

PROLOGUE

Dr. James Creed, Ph.D
St. John's College
Revere, Massachusetts

August 17, 1939

Dear James,

New York agrees with me. Unbelievable as it may seem, for the first time since Marie died, I am hopeful. I met Edward at the train yesterday, his aunt allowing him to travel the distance by himself, which made him feel quite the young man. He has grown so much in these eight months that I scarcely recognized him. My previous memories of him are dim, as my work kept me from the family so much, leaving Marie with the primary responsibility for his care, and it is a shock to recognize so much of myself, and her, in him. He is bright and eager, with wondering eyes that view the world as a miracle. If our work were not so important, I would abandon it, so that he could live with me.

Contrary to my fears, Edward bears me no malice for the arrangement, which he understands to be beyond either of our control. Indeed, he immediately drew me into his confidence by showing me the magazine he had brought with him, something called Astounding Tales. Science fiction, he tells me, and he describes it as the most wonderful literature in the world. His aunt disapproves of these pulp books, of course, calling them gaudy rubbish, and she will not allow them in the house. He saves his dimes to buy them, and stores them in the garage, where his aunt does not set foot. I found the magazine startling. It is indeed gaudy, and much of the writing in it will never qualify as literature. But the ideas! If nothing else, it looks always, always, to the future. In a better world, in the hands of better authors, what a literature it could be.

The better world, the future, has been much the topic of discussion these past two days. Edward could not wait to attend the fair. I confess that when he first wrote and suggested we visit there, my expectations were as low as my mood. I had visions of clowns and wrestlers and cheap little men slurring invitations to view all manner of fraudulent curiosities. Instead, this World's Fair is nothing short of amazing.

Edward was in awe of Manhattan, having forgotten a much earlier time when it was his home. Through the eyes of a child, I suppose. I feared the noise and frenzy of the city would frighten him, but he was energized! The island was a far cry from the last time I visited, half a decade ago, when the impoverished — not the congenitally derelict, but men and women formerly of station and grace — lined the streets, begging for whatever coins or crumbs might get them through another hour. While beggars are still widely evident, New Yorkers now move with purpose and industry, and you can see the future in their eyes.

Indeed, the entire city smacks of a garden. The sun has pleasantly shone the entire trip, imparting calm and well-being, and a steady breeze prevents any discomfort. Flushing Meadow Park, where the fair is held in the borough of Queens, is one of the most beautiful sights I have ever witnessed. We hired an automobile and driver to take us to the fair, though Edward was disappointed, for he had read of the train system and wished to ride it to the fairgrounds. From a distance, the fairground itself is most impressive. Unlike the county and state fairs which account for my entire experience of such things, all the buildings were specifically designed for the exposition. The main impression is light. The central buildings are painted a bright white, and the sunshine reflecting from them seems to turn the air to the purest glow as well. At night, these structures are lit and visible from many places in the area.

The greatest surprise is the sheer futurism of the entire event. Edward and I scrambled gleefully from exhibition to exhibition, and I felt much like a small boy myself. For there is tomorrow, set before us, as fantastic in shape as any film by Mr. Fritz Lang. While many of the exhibits are of a commercial nature, as one might expect, the grounds center around a giant dome, giving shape to as pleasant a civilization as has ever existed on this globe. Here, a model city has been built, on the design of a wheel, where a business district forms a hub, and homes radiate outward in all directions, homes for families, with fresh air and sunshine and lawns. Enough homes, it seems, for everyone. This exhibit, called the Democracity, is obviously designed with the human being in mind. As philosophies lately proliferate that object to the human being on principle, it is encouraging to see so massive an enterprise that accounts for every human need.

There is also a ride, known as the Futurama, that Edward and I greatly enjoyed. Elevated, it carries the rider over a scale map of this country, but it is a United States almost beyond imagining. The cities are tall and clean and proud, linked by snaking roadways large enough to handle as much traffic as might ever exist. Walkways rise above the streets, keeping pedestrians safe from accident. Farms and residences abut, with schools and recreational facilities at easy access for all regions. As in the Democracity, though there are differences in design, every human need is anticipated and met, with areas designated for commerce, for study and for play. This would truly be a world worth living in.

Or should I say all but one human need? I was struck by the absence of churches from this design. Was this an unconscious omission by the architects, or an assurance, after the depression and widespread human misery and rumblings of war from Europe and the Far East that have assaulted us for the last decade, that man's destiny rests in his own hands?

This question lingered in my mind as we approached perhaps the most impressive of all the exhibits, which I have dubbed the Golden Man. It was to this exhibit that we returned several times, as it was eerily reflective of the work we do, James: a robotic figure which speaks and moves through a series of tasks. Certainly, it could have been trickery, the last vestige of the carnival in this great exposition, but the Golden Man appeared more than a puppet on some hidden electronic string. In him was evident a new view of man, a concept of the human machine, which I mean in no way derogatorily. We have both suggested that man and machine become increasingly like each other as the Industrial Age proceeds. The Golden Man suggests that the machine must ultimately become man.

And man must become what? The machine? I do not think so. The machine must scruff away that detritus that holds man to the past, mainly crushing and protracted labor. The machine must make man free, or what use is it?

Edward and I returned to our hotel later than expected. He claimed a desire for real New York hot dogs and egg creams, and I obliged him, feeling none too hungry. The day had left me dizzy. I felt like a character in Edward's magazines. The future is here, James! It's here, and there are more men than we who have awaited it. In that future are the seeds of joy and a perfect world, without suffering, poverty or war. Perhaps, enough to satisfy even the German madman.

Edward and I will return to the fair tomorrow. How I wish he were returning to England with me, but we both know that is not practical, given the hours our progress demands. When I put him back on that train, my heart will break, but I know it is a better world I help to build for him. I am anxious to return to the laboratory. Our project seems no longer a folly by which I consume my waking hours, but work essential to the future happiness of the world.

Sincerely yours,

Benjamin Robillard

PART ONE: ELLIOT
Switzerland

Elliot Hecht stood in the driving rain and wished he were anywhere but Switzerland. Not, he reminded himself with a sniffle, that it was a bad place to be as the year rolled over into 1945. At least, it was unlikely the Swiss would ever shoot him as a spy. Others, who had trained with him, had been assigned to Germany, and, on restless nights when he couldn't sleep and on nights such as this when he made his solitary patrol of the northern border, he wondered which of them would come out alive and which had already died. He doubted he would ever know. They were warned not to exchange personal information during their O.S.S. training sessions, on the principle that the less they knew, the less they could tell if caught, and so he didn't even know their real names.

He took a sip of gin from a small, silver flask, to warm himself.

His own name during training had been Karl. It was the name he still used now. Karl Hagen, after the town in Western Germany. A hundred years earlier his great-grandfather had emigrated from that area to America, to flee conscription in the civil wars wracking the Fatherland at the time. With only the corrupted name of his destination to guide him and unable to speak a word of English, he had made it across the ocean and a quarter of America, before arriving on the doorstep of relatives in Decatur, Illinois.

By then, he considered himself not German but American and passed this status down to his children. The old man had still been alive when Elliot was born in the same big, white frame house in Decatur where his great-grandfather had settled. But the flu took him in 1918, and Elliot's grandfather, too, and left the baby Elliot with damaged lungs.

When the Japanese bombed Pearl Harbor in 1941 and America rushed to war, like most American boys, Elliot had tried to enlist. His lung problems kept him out, to his disappointment. It wasn't as if he were incapable; he had worked as a beat cop for the Decatur police department. But Decatur, wild enough during bootlegging days, had, by the end of the 30s, gone quiet. The war, the enlistment sergeant patiently explained, would be much more strenuous.

Swallowing his frustration, he had gone home to face the indignant stares of neighbors who

thought he was shirking his
duty, and the accusations of strangers who claimed he
was siding with his German cousins. By 1943, there were rumors of
German-Americans being rounded up to prevent spying, and when the
unknown man came to his door, Elliot was sure they had come for him.

They had. Where the Army had let him go, the Office of Strategic
Services, America's new spy service, had studied his file and decided to
recruit him.

So, here he was. Spying was the best way he had to serve his country,
and a spy Elliot became. Dropped by plane near Bern, he had walked eighty
miles or so to the province of Aargau, and, there, he'd set up shop across
the border from the legendary Black Forest.

He expected his unexplained presence would quickly draw attention to
him. To his surprise, no one seemed to care, and, ironically, the locals
assumed he was a German deserter. As war tore Europe apart, Switzerland
remained peaceful and remote, unconcerned. Within months, he was part of
the scenery, as if he had always lived there. He scanned the border at night,
watching for suspicious troop movements. He kept notebooks, listing
descriptions of every German who passed through town, and radioed the
information back to London, weekly. Once a month, money appeared in
his bank account. That was it—no assignments, no heroic raids on
Germany. Just this endless watching. It was as if they had sent him not
to war, but on an extended paid vacation.

Originally, he resented the lack of action, but he had come to appreciate his situation. If Germany were going to invade, they would have
done it years earlier. Now, the Reich was beaten; everyone knew it.
German deserters, once a trickle, had become a flood, as starving
men dropped their weapons and marched across the border to food
and freedom and to avoid being sent to slaughter on one of the shrink-
ing fronts. The Allies relentlessly pushed closer and closer to Berlin,
leaving the Swiss frontier the only available exit. At night, Elliot still
occasionally heard gunfire and screams from across the
river and imagined border guards mowing down deserters
who refused to halt, but those moments had become less
and less frequent. Were the guards deserting now?
He coughed hard and walked down a forest
path. The moon blazed so brightly it was
almost daylight. The cold and damp burned his
weak lungs. The winter had been unusually
warm, but a chill was coming in with the new year,
and the rain was turning to sleet.

Elliot wished he were home in bed. With the war so close to an end, he saw little value in continuing his work. But that morning, he'd gotten a letter from his "Uncle Gunter."

Elliot followed the procedures learned in training. Before opening the letter, he returned to his room, bolted the door, set out and lit a candle, and poured himself a cup of gin. He left the drink to one side and broke the envelope's seal. Spreading the message out on a table where, if interrupted, he could easily spill the gin and tip the flame onto the letter, he began to read it. If anyone tried to get in, the letter (and, most likely, his room) would be ashes before they could break down the door.

He read the letter three times, decoding it in his head, as he had been taught to do, and tried to make sense of it. The message instructed him to go to an old chalet and try to get inside. Elliot knew of the place; he had passed it many times while out hiking. Built next to an old coal mine, it had been deserted during the First World War when the coal ran out, as the war effort's demand for energy outstripped the region's ability to supply it. He had never seen anyone near the chalet.

Yet the message ordered careful investigation, with a warning of extreme danger. He didn't know what bothered him more: that there was real intelligence to be collected in his own backyard or that someone else had uncovered it.

He waited, as usual, until dark, slipped on his leather jacket with the flannel lining and the black, wool cap that hid his blond hair, and began the long trudge through the night forest. Now, the sleet had grown heavy enough to stick to his coat and melt through. Elliot was drenched and miserable. The trees grew thicker, masking out the moonlight, at just the point where the path turned to uneven rock. His lungs burned.

At last he saw the chalet, half-hidden between trees and the hill it was built next to. He left the path and circled slowly through the trees, his eyes always on the chalet. It matched the image in his memory, abandoned and dark, doors and windows boarded over, shingles fallen off. He couldn't imagine why he had been sent, what London could possibly hope to find.

On the building's north side, he found the clearing. A dozen or more trees were fallen, knocked away from their trunks, their bark at the break jagged and splintered. Lightning, he thought. In Illinois, he had seen storms and tornadoes do such damage but not so uniformly. The broken trees formed a cluster, and the damage to all was almost identical, something that didn't happen in nature. Besides, he knew, you didn't get tornadoes in Switzerland, and while it had been raining for weeks, there had been no violent storms. He leaned against a wrecked stump to catch his breath and wipe the sleet from his eyes, and then he saw the smoke.

It slithered from the chalet's chimney into the ice sky, just a sliver, white vanishing on white. He squinted and couldn't find it again. A delusion? Slowly, he circled the chalet again, and there it was, a white sliver, growing to a steady plume.

The sleet no longer mattered to Elliot. He circled one more time, judging which side of the chalet was darkest. If the Germans were there, he had no urge to meet them head on. He stalked to the edge of the woods. The chalet showed no sign of life, but he imagined hidden guards, ready to shoot on sight. His legs were frozen. Now that he finally faced the prospect of heroism, he couldn't bring himself to run that fatal distance.

The wind solved his dilemma, loudly slapping a long pine branch against the chalet roof. Anyone inside would likely be used to that sound, and noises from the roof might not alarm them. Quickly, he found the tree, climbed it and shinned down the branch, praying it would hold his weight. At last his shoulders touched the roof, and he grabbed it fiercely with numb fingers.

As he had hoped, where shingles had fallen from the roof, the surface beneath was soggy. Elliot dug his fingers in and pulled away clumps of it, trying to block the rain from the widening hole with his body. He heard a soft hum and felt a tingle, but no one reacted inside. When the hole was large enough, he crawled through it.

He swung to a rafter and reconnoitered. The chalet had begun centuries earlier as an eating hall for miners, built in the architecture of the time: large, open rooms and squared beams. A soft electrical glow lit the room, and Elliot realized the windows had been blacked where they weren't boarded over. Alone in the room, just inside the door, sitting on an old oak chair, was a guard, a German soldier hugging a Schmeisser machine gun. He was half-asleep. Elliot slid along the rafter until he was over the soldier's head, then dropped to the floor behind him. The sound started the soldier, but before he could react, Elliot had his arm around the man's jaw and, as he had been taught, he broke the soldier's neck.

It was the first time he had killed. The only time, Elliot hoped. Trembling, he dragged the body behind a heavy wooden chest and hid it. Then he snatched up the Schmeisser, ready for others to come to the guard's aid. No one came. The place was deserted but for him and the dead man, and the hum tingled through him, a dull vibration rising from the floor.

The mine.

He felt his way along the floors and walls, checking for any sign of a door: inconsistent temperature, unexpected air current. Nothing. The chalet would not give up its secrets.

London, he knew, would not settle for that answer.

He closed his eyes and concentrated on the hum. Here it was louder, there softer. Elliot criss-crossed the room five times, pinpointing

the hum's center. When he opened his eyes, he stood in front of the fireplace, which was cold. The smoke rising from the chimney outside could not have come from the fireplace.

Elliot shook with excitement. After all these months, he knew this was the moment he had trained for, the moment when he would face the enemy. His hands ran over the gray stone of the fireplace. His thumb caught an edge, and it slid to one side.

The fireplace rolled off the wall. Behind it, stairs led down to an infernal glow. Electric lights, Elliot knew, but every superstition planted in him in church sprouted at that moment, and he expected devils to burst out at him and drag him down forever. He gagged on a thick smell of burned oil that burped at him from below. He wanted to run.

His finger on the Schmeisser's trigger, he stepped cautiously down the stairs. If this was hell, it was run by Germans. A handful of soldiers stood in a far corner, lighting cigarettes. Giant electrical cables snaked everywhere across the floor, connecting a maze of equipment and machinery, all dwarfed by a giant iron construct in the center. White-coated men darted back and forth, checking The Machine.

It was a complex of machines, but to Elliot, it was The Machine, a monster spewing sparks and gasses, as tubes everywhere pulsed an unearthly red, and it roared its incessant electric whine.

Blood pounded in his head. He couldn't believe they didn't hear it, or could stand it if they did. He slid under the handrail, and beneath the stairs for cover, and stared at the awful scene. The cavern trailed off into darkness. In the distance, an intense white light clicked off, and he could hear muffled voices cursing in frustration. As he hid, a chubby man in a surgeon's gown stormed by, covered in blood. In the man's soft cursing, Elliot could make out only one word: Grosswald.

Two soldiers followed close behind, carrying a body. An old man, terribly mutilated, with numbers tattoed on his wrist. They dumped the cadaver on a rockpile in the darkness, and went back to join the others. Elliot held his breath and listened, but all the voices were far away. Sticking to the shadows, he crept from his hiding place for a better look at the corpse.

He gagged as a sour heat rocketed up his throat. The pile on which the old man had been dumped was not rocks but bodies: men, women, children. All had two things in common with the old man. Each, had an identifying number on one arm, and every cranium, and the brain within, had been removed.

The cool of the Schmeisser burned his hands. With mad eyes he stared at the soldiers and the technicians, and, in his head, he saw them fall, blood spitting from their chests, their heads, and his finger on the trigger of the gun that killed them. A vengeful hatred engulfed him. In that moment, he was not Elliot Hecht, nor Karl Hagen, nor was he an Allied agent, nor even a man. He was the hand of God, come to hell to deal hell to devils.

And then the moment was broken.

He felt the presence before he saw it and spun, howling, firing the Schmeisser at it at point-blank range. As the gun thundered in his ears, and sanity crept back to warn him that his cover was blown, Elliot felt the sting of a bullet graze his arm. At first, he thought someone had shot him, but he saw no gun in the figure's hand, and it had all happened too quickly for the others to react. A Nazi officer stood before him, almost eight-feet tall, and he realized two things at the same time: the officer he'd just shot, in the crisp dress uniform of an SS Major, had not fallen in the gunfire, and his skin was black. Not black like the Blacks he had known in Illinois, but polished black and shiny. Elliot fired again, and, again, the bullets ricocheted away with a sharp whine.

The Major was made of iron.

The Major's arm whistled toward him, faster than was humanly possible, and Elliot was hurled back, slamming into the rough stone floor. Somehow, he managed to hold onto the Schmeisser. Now the soldiers rushed for him, but the Iron Major waved them off before they could shoot. "Alive!" the Major ordered, in a hollow, icy voice. He spoke

in German, but even in the haze Elliot crawled through, he understood the meaning. They would take him alive, for whatever reason, and put a number on his wrist. His brain would be removed, and his body dumped, to be forgotten like all the others.

His finger twitched as he rolled, and the Schmeisser sang again, not for vengeance but for survival. The soldiers and technicians fell back, and the monster started for Elliot. Everything in him was on fire, his nerves and his lungs, the gouge in his arm, but he found his feet and made them move. The rest was a blur. The stairs, the chalet, the night, he saw none of them. The Schmeisser slipped from his hand, clattering on rocks, but he kept going. There was no path, but he kept going. He ran and ran until he could no longer feel his legs, then toppled down an embankment. Elliot came to rest on a riverbank, his head half in the water. He gazed at the infinite stars above him and thought of nothing.

England

*I*an McBane heard his name called and rolled away from the noise, trying to find peaceful darkness. He had been working steadily for eighteen hours, sorting through reports until his vision was too blurred to go on. Nothing new had come from his efforts. All intelligence indicated a straight-line march of Allied troops on Berlin, with few elements of the shattered German army to bar their way, and civilians everywhere welcoming them as liberators. Total victory in Europe, by his calculations, would come in early June at the latest.

"McBane!"

It was Timothy Severn. As McBane placed the Englishman's voice, he quit all thoughts of sleep. Severn was the product of good British breeding, as the sergeant at the gate put it, and unlikely to disturb a man without cause.

"McBane, get up. There's a wire for Uncle Gunter."

Gunter, that meant Karl. Trying to rouse himself, he worked his way through the maze of rules and deceptions he had devised. Karl, his agent in Aargau. Lucky Karl, left out of the war. McBane had agents spread throughout Europe, at crucial points, and his contact name was different for each. It was how he kept track of them. Only Karl would have sent a message to Gunter.

Dread rippled through him. McBane had never known Karl's real name, that was another of his rules. But Karl's assignment had been to investigate something that could possibly thwart an Allied victory and give Germany an unbeatable advantage. While McBane rarely shared it with others, he had a recurring nightmare of the Third Reich playing one last horrible trump even as the world thought the war was over. He saw every new message as a harbinger of that surprise that he knew he, maybe the whole world, would not survive.

He reviewed Karl's mission. McBane had tracked the caravan for weeks, a procession of vans and soldiers that had fled Castle Grosswald in Germany in advance of the approaching Soviets. The caravan wound through the Reich, avoiding towns and any likelihood of combat, before disappearing just north of the Rhine. He had sent Karl to look for them.

The memory jolted him awake. Severn stood above him, holding out a piece of paper. The Englishman seemed impossibly young, McBane couldn't recall ever being that young. He couldn't even recall a time before military intelligence. It had eaten up his life. Severn probably had a girl off base, and a dream of a post-war future. All McBane had was the work.

He read the coded message, and the room seemed to drain of air.

"Grosswald," he muttered under his breath.

"You're sweating," Severn said. "Shall I call a doctor?"

"Call a staff meeting." McBane sat on the cot and checked his watch. The dim overhead bulb flickered. Electricity was so unreliable in England these days. "Twenty minutes."

"Twenty minutes? I don't think everyone's here."

"Call them. Mandatory attendance, no exceptions. And get me a line to the Ministry." He took a long breath, glancing into Severn's stricken face. "All right, thirty minutes. I need to clear my head."

>×<

The drizzle of English winter was colder than any snow he could remember. It soaked through everything, even weighing down the mist that filtered from his nose and mouth as he exhaled. With the temperature hovering at 40 degrees night and day, he couldn't shake the chill. He hadn't seen the sun in weeks. It depressed him, it seemed a dire omen.

Especially now.

Schoolbuses rolled to a stop at the door of Hembley Manor, and McBane stopped to watch that day's Coding Squad file off. The girls were secretaries from town, girls with skills at working out the cryptograms, jumbles, and crossword puzzles in the London papers. A hall in the manor had been divided into cubicles, and every day, the Coding Squad — the number and personnel changed each day, another security measure — sat one to a cubicle, deciphering a single word of intercepted German code. Not one of them would see two consecutive words, and rarely more than one word from each message, and none would ever know exactly what they were working on. Hundreds of girls in numbing work, changing the course of the war and never knowing how.

Real troopers.

A couple of girls noticed him and smiled before vanishing inside. Hembley Manor was eighty miles northwest of London, on a fourteen square mile estate, and as far as McBane could tell it had hardly been retouched in the last 200 years. It was a drafty, dour place, but it was beyond the range of German rockets. In earlier days, the secretaries had been thrilled to come here and get away from the noise and collapse and constant unexpected bombings in the city. They still enjoyed the countryside, but with the best launching sites on the French Coast recaptured by the Allies, German bombs rarely fell on England anymore, and the girls were notably cheerier these days, and already considered the war as good as over.

McBane wished desperately to share their confidence, but he knew things they didn't.

He thought of the spy caught a month earlier, searching for the miracle machine that had cracked the German codes. If the poor man only knew the Nazi High Command had been outsmarted by shopgirls and secretaries, and conned by a lie.

McBane was a professional liar now, a mythmaker in service to the war effort: a spymaster. Even his friends no longer trusted him. Only those, like Karl, who had never met him still held faith. Poor doomed Karl, whom he would soon betray.

"McBane!" Severn again, calling from a second floor window. "The Ministry's rung back for you."

"Be right there," McBane shouted, and he hurried back inside.

>-><-<

By the time he reached the conference room, the others had gathered there. James Creed tapped his wristwatch impatiently. "We were led to believe this meeting was urgent." Creed was an aging aristocrat, appalled by McBane's American nationality and Scots-Irish ancestry, and he missed no opportunity to let McBane know it. McBane loathed the man. He didn't know how Robillard, Creed's partner, could stand him.

"Sorry I'm late. I was on the line with the P.M."

"Churchill?" That shut Creed up.

McBane took his seat at the head of the table, Severn to his left, a white-haired secretary named Jenny to his right. She switched on a wire recorder as he began to speak, a record of the meeting for official files.

"I've brought the P.M. up to speed on the situation, and he has promised full cooperation."

"Situation?" asked Robillard. A neurologist by trade, he had the unique distinction of being the first boy born in 1901, and Creed loved saying Robillard was a child of the century. McBane didn't think that was far from wrong. Over the past decade, Robillard had left medicine and turned his skills to a pursuit usually

relegated to B-movies and pulp magazines: the creation of an artificial man. "You're a spy, McBane. What possible use could Creed and I be?"

"Do you remember Uwe Grosswald?"

Robillard shrugged, but Creed said, "Met him once at a conference in Amsterdam. What was it, 1929? Revolting little madman, innocuous enough in a crowd but let him catch you to one side, it was all talk of eugenics and the coming race and all that Nazi rot. Died a few years back, didn't he?"

McBane nodded. "1937, February. I lost one of my men to him."

"He was under investigation?"

"Dr. Creed, you do know what a robot is?"

Creed blanched. "Of course, we do. What do you think we've been working on?"

"Robot," said Kathleen Reay. A Scottish-born beauty, she worked as a lab assistant to Creed and Robillard, and spoke in a solemn tone that demanded to be taken seriously. She gazed over the room, pausing for just a flash as her eyes met Severn's, and they traded smiles. Severn looked away, reddening, but McBane caught it all, biting back a smile of his own.

Timmy Severn, he thought, you do surprise me.

"A mechanical man," Kathleen continued. "The term was first coined by Czech writer Karel Capek, who used them as a metaphor for the dehumanization of man in the industrial age. In a 1921 play, I believe. *R.U.R.* Rossum's Universal Robots."

Someone in the room whispered, "Robillard's universal robots," and a snicker spread through the room until an angry glance from Creed silenced it.

"That's where it started?" McBane asked. He hadn't heard the story before.

"The name was attached then," she continued. "But the concept goes back almost as far as man. Greek mythology discusses artificial men, and legend has it DaVinci tried to design one. Pope Sylvester II was thought to consort with devils because he allegedly had a mechanical head, a gift from unknown beings in the far East, that answered yes or no to any question put to it."

"And what's this got to *do* with anything?" Creed demanded. "Next you'll be bringing up homonculi!"

"Tiny men created in alchemists' laboratories," McBane said.

Creed stared at him. "A well-read cop. How interesting."

"I'm not a cop," McBane replied. "Some background seemed appropriate. Did I mention Grosswald built a robot?"

"Functioning? It's impossible."

"I assure you, Doctor, it's not. It killed him. Then it rescued two teenage children before being destroyed."

"Impossible," Creed repeated.

"Dr. Robillard? You've been quiet. What's your opinion?"

At the far end of the table, Robillard twirled the end of his beard and thought a moment. Forlornly, he shook his head. "I agree. It's impossible."

"Why?"

The mechanics of it, mainly," Robillard said. "We tend not to appreciate what a marvelous piece of machinery the human brain is. To work systems as complex as ours, thousands of commands must be processed every second. Grip the edge of the table there. Both hands."

McBane did.

"A robot would need a packet of commands half the size of this room to accomplish the same thing. It would need a compact storage system and a packet retrieval system that would function on command. Frankly, it's beyond human technology, for now."

"There's another one," McBane said.

"What?"

McBane smoothed the paper out in front of him. "Uncle Gunter. Stop. Investigated as requested. Stop. Building cover for laboratory comma Nazis involved. Stop. Almost captured. Stop. Encountered man of iron comma bullets bounced off him. Stop. He is exceptionally strong. Stop. It goes on in some detail, but that's the gist of it."

Creed smirked. "Uncle Gunter?"

"You didn't hear that. Did my man describe a robot?"

"Grosswald died," Robillard said. "How could he create a new robot?"

"My man describes someone who sounds like Heinrich Muller, Grosswald's assistant. My guess is Muller continued after Grosswald's death, and has been doing his bit for Nazi science."

"Nazi science," Creed muttered. "What a joke. These people believe in a hollow earth, for heaven's sake. They claim the stars are blocks of ice, hovering in space and reflecting sunlight. They're insane."

"They created the V-2 rocket," McBane reminded him. "Would you care to see an army of iron monsters tramping through Piccadilly Circus?"

"What do you want us to do?" Robillard asked.

"Golden Boy," said McBane, but he knew Robillard already knew, because, in the end, what else could it be?

"No," Robillard said. "He's not functional. He's a party trick."

"He moves by himself."

"But he can't think by himself. He needs someone calling the shots."

"My man can do that."

"Think it's that simple?" Creed said. "Why don't you send in an army to wipe this out, what-ever it is, if it's so important?"

"Switzerland is neutral territory, Doctor. I'll try to get their government's permission, but inva-sion is an act of war. I doubt Churchill or Eisenhower would look kindly on that. How much time will it take to prepare Golden Boy for duty?"

"Fully operative?" Robillard said. "A rough guess — thirty seven years."

"I can give you eighteen hours," said McBane. He stood and started for the door. "Do your best. I'll contact my man."

Switzerland

*E*lliot read the message a third time. It made no sense.

Veins throbbed in his forehead. Since the night at the chalet, he suffered from migraines and terror dreams. He could scarcely close his eyes without seeing that face, sculpted and polished into a mask of scorn, infernal in the torchlight. Other faces joined the Iron Major's in his nightmares: the mutilated faces of the dead he had failed to save, their skulls broken, their brains gone, their empty faces pleading for justice. Sometimes he saw his own face there, his body lost to darkness, and then it was not his face but a new face of iron, screaming his silent scream.

To his surprise, they hadn't come after him. He presumed they didn't know who he was, or where, but it was just as likely they were unwilling to risk exposure. What could he tell the local authorities? A giant, iron German threatened them? Elliot imagined being packed off to an asylum—even to him the story sounded insane. So, it was a stand-off, and he quietly wished they would slip away in the night, back across the frontier, as secretively as they'd arrived, saving him from facing the Iron Major again.

"Uncle Gunter's" message dashed that hope. Decoded, it spelled out the most desperate plan. He downed his drink and steeled himself for the mission. At once the dread crept back, and he had the delusion of eyes on him, but when he glanced around the small café, no one seemed to pay any attention to him.

"Another?" asked the waiter, in guttural German, and he reached for the glass, but Elliot waved him away, dropping a handful of francs on the table to cover his tab. Scanning again for watching eyes, he went to the café's washroom, lit a match and burned the message to ashes in the sink.

To reach the meadow, Elliot had been forced to drive around the rim of the lake, left wheels splashing the water. Renting the truck, a large flatbed with under-inflated tires and no roof, was difficult, but reaching the clearing had proved to be nearly impossible. He sat in the dusk on the lip of the truck bed and watched the western sky. Snow dusted the meadow, except where the tires had churned up mud, but no sound disturbed the stillness.

Elliot cursed the farmer who had taken his last franc for the truck, and cursed London. He was their man in the region. Why had they chosen such an inaccessible site, instead of asking him to name a drop point?

Not that the money mattered. He didn't expect he'd need it.

Dusk turned to night, and twice Elliot stopped himself from drifting off. There was no heater in the truck, and he wished he had brought the gin flask. He could use the warmth. He felt his fingers and toes begin to stiffen.

From the distance, thunder rolled, and as Elliot listened, the dull rumble opened to a staccato thumping, high above. Planes in night reconnaissance over Germany, swinging wide and south as they turned back, straying over the border. Elliot switched his headlights on. The roar dipped low.

As Elliot scanned the sky, he saw the planes, black crosses against the moonless indigo night: two fighters escorting a larger cargo plane. The cargo carrier flew still lower, almost touching the tree tops.

Then twin, white parachutes flowered in mid-air, and the planes soared off as the blossoms floated downward. Elliot saw the package, dangling under the parachutes, and it looked to him like a giant's coffin. He started up the truck and rolled across the meadow. The wind had risen, swirling the dry snow into meta-morphing patterns, and if it caught the parachutes with enough force, the package might be dashed into the mountains or blown off-course into the lake. Either would bring Elliot's mission to a halt.

Part of him prayed for that; it would simplify things. He hated the thought of meeting the iron monster again. Ashamed, he remembered running madly from the cave, though he had not shared the memory with London. He could still run; he had an identity apart from his real self, and it would be easy to vanish in the chaos that would be Europe at war's end. A spy, trained in such things, could get away with that.

But Elliot considered himself a soldier.

The crate settled to earth, falling on its side under the parachute shroud. Elliot pulled along-side. For several minutes, he tried without success to budge the crate.

He was starting to lose patience with London. What had they been thinking?

He pulled a tire iron from the back of the truck and worked on the crate lid. Nails loosened and boards began to split, but it was too slow. He couldn't feel his fingers. Frustrated, he smashed at the planks with the iron.

As splinters flew, gold glinted from inside the crate. Elliot dropped the iron and pulled at the boards with his hands, working them back and forth until they came free.

He gaped. It was a joke, it had to be. He understood, then; London had thought his report a prank or a delusion, and played a joke on him. It was the only way the night made sense.

Inside the crate was a golden statue.

The rifle shot punched a hole in his arm. Elliot fell, in shock, and from the woods all around he saw dark figures moving. They were following me after all, he thought, and he tried to stretch for the fallen tire iron, the only weapon at hand, but his arm flopped at his side. German soldiers closed in on him. A rifle butt slammed against the side of his head, and Elliot pitched into a pure darkness that held no iron faces.

><><

Something wet and rough scraped his cheek, and Elliot opened his eyes.

His head rested in the lap of a woman who stared down at him with sad, hollow eyes. Those eyes said she was young, no older than he, but her skin was drawn so tightly on her face that she looked like an old woman. Her head was shaved clean. With a skeletal hand, she ran a rough cloth over his face and sang what sounded like a children's song in a language he didn't quite understand.

"What are you?" he said, in German, and in perfect German she whispered:

"The walking dead."

He sat up. Blinding pain pulsed through his wounded arm, and he let out a yelp before she clapped that skeleton's hand over his mouth.

"Shhh," she said. "They'll hear."

She stuck the cloth in his mouth to muffle him, and he bit into it as he got to his feet. Someone had bandaged his wound while he was out. Gently, the woman took his hand in hers and helped him flex his arm back and forth until he grew used to the pain. The wound still burned, but he could ignore it.

As his sight returned, Elliot pulled the gag from his mouth and gazed around. They were in a cave outcropping, and wooden crossbars had been tied in place to turn the area into a cell. Now he saw other faces, eyeing him fearfully: the walking dead — men, women and children, as emaciated and drawn as the woman was, all their heads shaved.

All with numbers on their wrists.

"What's your name?" Elliot asked, and she turned away, staring at nothing with bitter eyes.

"You took everything from us. Parents, children, lovers. You butchered the world and made this of us. You taught there are things worse than death. Why? What name would I have left?"

"I did?"

"You are German, no?"

He began to shake his head, but stopped himself, remembering his training. Ashamed, he nodded, and she mistook his shame for guilt. Scorn lit her dead eyes.

"I didn't…" he began. "I didn't know." Her scorn only grew. "Who were you, that they did this?"

He wanted her to say they were criminals, or revolutionaries, anything that might let their punishment be somehow justified, but she said, simply, "We were not them."

Elliot sucked in a small cry and began to rock back and forth, trying to catch his breath. The woman drew him to her, cradling him, and said, "Elka."

He knew it was her name. "Karl," he replied, a reflex, and the lie tasted bitter on his tongue. Her eyes narrowed with mistrust; hovering on the edge of life and death, she had no time for lies. Still she held him, consoling him, a press of leather and bone against his skin.

"You comfort me, when you hate me?" he asked.

She said, "We are not you."

><<<

The soldiers came for Elliot as he slept. Roughly, they lifted him to his feet and shoved him outside, as the other prisoners turned away. He looked for Elka, but she wasn't there.

He had expected monsters, but the soldiers were just boys. The woman had been right, he was them, indistinguishable. In their faces was a hardness built on fear. They were afraid, too, but not of him.

The pens, as the soldiers called them, were deeper in the mines than he had seen before, and they marched past what looked like assembly lines, with no conveyors moving. Wrought-metal limbs and heads were scattered everywhere, amidst blacksmith tools from another century. Medieval and modern mingled haphazardly in the cave.

They marched past what appeared to be huge refrigerators and the large, unknown machines that whirred and clicked as he neared. The cave stank of gasoline fumes, and Elliot could hear motors somewhere. A soldier shoved him into a wooden chair, and the others fell back, moving off to other duties as two remained behind him, waiting.

From the darkness, a clanking gait approached. A bright light hit Elliot's eyes, and he squinted, making out the Iron Major just beyond the light. Behind the Major stood the chubby surgeon, something metal glinting in his hand.

"Name," the Iron Major demanded.

"Karl Hagen."

An iron hand clamped onto Elliot's skull and slowly squeezed until Elliot thought his head would burst. He screamed. The Iron Major let go.

"Name," the Iron Major demanded.

Elka had been right, there were things worse than death. He had dreamed of being a soldier, a hero, a great man, but now, he was just a man, broken by pain too intense to bear and willing to do whatever was necessary to spare himself more pain.

"Elliot Hecht," he said, surrendering.

"English?"

"American."

The surgeon moved forward, his smock blood-spattered. The scalpel in his hand was lined in red. "I am Heinrich Muller. Your superiors briefed you about me, yes?"

"No," Elliot whimpered.

Muller seemed disappointed. "Tell us about your robot."

"What?" Elliot said.

"Your mechanical man." The surgeon had already lost patience. "The golden man in the box."

"Robot?"

The Iron Major tapped Elliot's face, and he felt his cheekbone fracture. At a wave from Muller, the soldiers pulled Elliot from the chair and dragged him across the floor.

They dumped him onto the opened crate, which stood against a wall, as Muller and the Iron Major followed. The golden man was inert, useless.

"A wonder," Muller said reverently. "Four hundred years ago, the great anatomist Vesalius mapped the human system, unearthing the secrets of blood vessel and synapse, of nerve cord and brain. He, too, worked directly from raw materials."

"People," Elliot said.

"Corpses. It put him at a disadvantage. There are subtleties of life that cannot be revealed in death. I eliminate that deficiency, and so those subtleties can be replicated." Proudly, Muller slapped a hand on the Iron Major, then turned his attention to the robot in the crate.

"It seems a shame to tear it apart," he continued. "Such creatures are art. But I should like to know how it works. Perhaps you could tell me?"

Elliot shook his head. Muller shrugged and studied the golden man.

"We had a design like this once. My mentor, Dr. Grosswald, called it a good design, but we discarded it. The nervous system was deftly imitated, simple wiring and mechanicals, but the body—it could be made to move, you see, a handful of commands, and that was the extent of a robot's abilities. Hardly useful for sophisticated purposes, eh?"

"Hardly," Elliot replied, though he no longer listened.

"The solution," Muller said, "was obvious in even the first literature on the subject, and we laughed that we had not seen it sooner. We needed a storage vessel, capable of holding and processing a nearly infinite number of commands, and quick enough to sort and initiate them as necessary. A medium capable of independent, albeit controllable, thought.

"But there was no brain, we learned, better than a brain."

"A brain," Elliot murmured back, and it was only an instant later he realized what he had said. The dark face of the Iron Major hovered over him, and he knew, at last, what the metal skull housed. All the dead, all the mutilations, came rushing back to him with new meaning.

"Of course," Muller continued, resignedly, "the personality was a problem. Dr. Grosswald was the Vesalius of the brain, mapping the regions of control. A few surgical motions and snip-snip! The physical seat of the personality is expunged."

"And only the machine is left," Elliot said.

"A slave to the machine, but that is the future of man. Flesh has not the durability of metal, sad to say. Like batteries, the brains burn out, to be discarded and replaced. The Major requires a steady supply, and he especially enjoys the piquancy of his enemies' brains." Muller patted Elliot's cheek in mock sympathy.

"Cheer up. It will be a useful life, while it lasts. The Major especially appreciates the enforced service of his enemies. It adds such piquancy to the experience. He does like his little trophies."

Then Muller stepped close, his lips brushing Elliot's ear, and in a seductive whisper that held out a last desperate chance of survival, he said, "Tell me about your robot."

Defeat had clouded Elliot's thinking, but confusion snapped him alert. A lifeline from the surgeon? For an instant, he felt new desolation; he knew nothing to tell them. In that instant, he ran Muller's words over and over in his head, searching for a scrap to throw them.

For the first time, he listened, and understood he was wrong.

He knew three things.

He knew what was evil.

He knew character; it was what had convinced them to recruit him. Muller's offer was out of character and unnecessary—if the surgeon were telling the truth.

"You can't do it," Elliot said. A statement, without fear. Muller's smirk vanished. "That's why all the parts, and only the Major. You want an army, but he's the only one you could make."

"No," said Muller, without conviction.

"You don't need me. You're the scientist. Tear Golden Boy apart if you want his secrets." Muller stammered for a reply. Elliot smiled. "You can't risk ruining him, can you? This place, all those people, all the dead—you're trying to build another Iron Major—and you can't do it! You are a failure!"

With a howl, Muller lunged for him. Elliot sidestepped and shoved him away. The soldiers and the Iron Major were already responding, but time froze for Elliot. It seemed he was swimming through ice, darting for Golden Boy, and his mind burned with what he knew.

He knew The Word, the meaningless word, spelled phonetically in the message from London.

Elliot felt a searing sting at his neck, and another roaring insect punched through his side and burrowed deep. Bullets, he realized, but by then he had fallen into Golden Boy, clinging to the metal man. His lung ached as he murmured The Word, then he was weightless, lifted high into the air.

Smoke and light blurred as he slammed onto the floor. Something snapped in his chest. The Iron Major let him go and raised a hammer of a fist above its head. Elliot heard shots and Muller's screams, but they were somewhere else. The whole world for him was that plunging fist and the long, last breath that tore his lungs.

Golden Boy's hand jutted between them, an impassable anvil to stop the Iron Major's hammer fist.

Elliot rolled away as Golden Boy and the Iron Major locked together, swinging at each other like drunken prizefighters. Their metal boots pounded the floor. Soldiers raced toward the fighters. Shots flattened against both robots and left them undamaged, but the soldiers kept firing, over Muller's frantic protests. No one came after Elliot; in the chaos they had forgotten him, and he pulled himself out of sight to rest, while metal rang against metal.

Blood soaked his shirt. Bone stuck out from his side to poke his elbow when he held his arm close. Tentatively, he dabbed at his face and chest, to find both ruined, and he almost passed out at his own touch. It was then, with rare and absolute clarity, that Elliot knew a fourth thing: he was going to die here.

As he pulled himself to his feet, Elliot witnessed the soldiers wasting ammunition on the battling robots. The bullets still had no effect. Something bigger would be needed to stop them. The Iron Major stepped backward, crushing a soldier underfoot, and the others scattered. The robot's fist crashed against Golden Boy, knocking it aside. From their new position, the soldiers opened fire again.

Elliot imagined a platoon of the monsters, retaking half of Europe.

The robots plowed through Muller's machines as they fought, shattering them, and above the pounding, Muller's incoherent shouts rang out. A soldier darted between the robots and shot at Golden Boy. Golden Boy's fist swung toward him.

And stopped, inches from the soldier's head.

Muller howled with delight. "It won't hurt a man! It has a weakness—they programmed it to spare human life!"

The Iron Major tore the soldier's head off as it slapped him aside and drove a jackhammer punch at the other robot. Golden Boy slapped the hand away. "Stay out of it," The Iron Major rasped.

"Uwe! Wait!" Muller shouted. "Think!"

Uwe? thought Elliot. He scurried from cover, grabbing a Steyr-Solothurn machine gun from a crushed soldier. At Muller's command, more soldiers formed a line between the robots. Golden Boy froze, unable to fight them.

Elliot mowed them down.

Muller ran, deeper into the cave. The surviving soldiers ducked for cover, shooting at Elliot, and Elliot hobbled toward safety, laying down cover fire. The Iron Major blocked his path.

"Die," the robot said.

"Keep your shorts on," Elliot replied, and shot the Iron Major in the eyes.

The shots were as harmless as any other, but The Major covered its face, staggered. A human reaction, Elliot guessed, as he darted past it. He had formed a theory, based on what he had heard, that what moved the Iron Major was not just electricity and wires, but a human consciousness or something based on it.

Muller was in sight, ripping at the large clicking machines. His back was turned. Elliot moved in slowly, training the Steyr-Solothurn on the surgeon.

"End of the road," Elliot gasped. His throat was dry, he barely had voice left.

Muller turned, eyes wide with
fear, arms filled with wire coils.

"My life's work," Muller said, offering
up the tapes. "After the war, my secrets will
make you rich. Wire recordings, all my data.
Take it, and let me go. Please."

"The war's already over."

"It isn't! There's still time!"

"Not for us."

Elliot fired. Muller's arm went limp and began to bleed.

The chubby man dropped with a moan, the coils skittering
across the floor. He dabbed his fingers in the wound and looked at his
blood, as if it belonged to a lab animal. "My secrets…" he started again.

"Your secrets," Elliot said. "How to destroy lives and spread misery. For
what? Even your robot couldn't escape humanity."

"Grosswald, Grosswald the madman…brilliant…"

Muller stuttered and slumped over.

With the rifle as a crutch, Elliot hobbled back to the pens, dragging Muller
behind him. The prisoners turned away as he approached. Most sat where he had
last seen them, unmoved by the commotion.

He flung the gate open. "Get out," he told them. Several looked at him, finally, and
he saw the distrust and fear in their hollow eyes. The walking dead, Elka had said.
"Walk. There are dead guards. Take their guns. If you don't leave this place now, you'll
die with it. Do you understand? Go."

Tentatively, they rose, watching him as if he were torturer, not a liberator. Slowly, they
filed past him.

Elliot locked Muller inside the cage and left him, ignoring his pathetic pleas. This was a
place of horror. He wanted it gone, all memory of it destroyed and buried. If it remained,
others would come, unearthing Muller's loathsome secrets. They would die here with him or
poison the future.

He passed the refrigerators and opened one. It housed row upon row of brains, carefully
sealed in liquid-filled jars, a case history pasted on each one. On one jar, he found a name:
Drezdenko, Elka.

He threw up.

Somewhere in the darkness were the motors, spitting out their hum and the stench of
burned oil. Elliot staggered toward the sound. He found old turbines, attached to huge pres-
surized fuel tanks. A soldier, a boy in his teens, trembled, and aimed a Walther pistol as Elliot
approached.

Elliot aimed at the fuel tanks. Stepping closer, he held out his hand. Sullenly, the boy put
his pistol in it.

"Go home," Elliot said.

A black streak glinted in the torchlight, slicing through the boy's skull.
Elliot pointed the Walther at the Iron Major's eyes.

"What happened, Grosswald?" he asked.

The Iron Major stopped. "How did you…?"

"You *were* a madman, weren't you? You chose to end up like that?" Elliot circled slowly, staying out of the robot's reach.

"Experiments go wrong," the monster rasped. "But I am the first of the coming race. A master race, without limits. I alone have pierced the veil of life and death. I recorded the soul, my soul, to be rehoused when my body expired. I created a machine that thinks my thoughts, and I will create legions in my own image!"

"But something went wrong."

"Muller," it said in disgust. "Always Muller."

"You're not Grosswald," Elliot said. "You're just a machine."

"I will live forever!"

"Nobody lives forever." Elliot pointed the Walther at the Iron Major and imagined a smirk on the immobile face.

This, Elliot knew now, was what he had been born for. London had known it. Golden Boy was a trap. They had meant for Elliot to be captured, to be questioned. To die.

Because this thing had to be done.

He pulled the trigger. The bullet missed the Major by several feet, piercing the fuel tank instead.

The turbines went up in a deafening torrent of flame, a wind strong enough to hurl even the Iron Major off its feet. The robot slammed into a cave wall. Explosions rippled through the installation in a chain reaction. As the power died, the lights went out, leaving Elliot bathed in the glow of hellish flames and unbearable heat.

The cave floor shook and spat as old tunnels caved in, and the Iron Major vanished in a cascade of rocks and dust.

Elliot reeled away, through the flames and smoke. Half-blind, coughing up blood, he staggered along the walls, trying to remember the way out. At last, he found the refrigerators. Like him, they burned. He passed the empty pens, and hoped the prisoners had all found their way out.

Another shockwave ripped through the cave, knocking Elliot down. His legs no longer worked. He dropped the weapons and pulled himself along, but the effort stabbed at his injured lung. He collapsed.

His hand hit metal. He lifted his eyes. A foot.

Gently, Golden Boy picked him up and carried him toward the exit.

"Stop," Elliot commanded as they passed twisted remnants of Muller's research. The fire had not spread this far and wouldn't. Muller's wire coils were still on the floor. With his last bit of strength, Elliot picked them up.

With Elliot in his arms, Golden Boy stepped into the Swiss dawn. For five miles they walked, until they were well away from the cave. Barely able to speak, Elliot said The Word, and the robot halted. "Set me down."

Golden Boy lay Elliot on the grass.

"I wish I were going home," Elliot said. "I'll be buried here. No one will ever know what happened to me."

Golden Boy stood over him, still.

"Hide the coils. Keep them with you. Never let anyone get them." Lights blinked on the robot's chest as it processed the command. Elliot's voice faded. He was blind.

"Come here."

Golden Boy knelt, its head close to Elliot's mouth. Elliot's last commands were barely a croak, but the lights on the robot's chest blinked at Elliot. He closed his mouth. Golden Boy marched into the dark of the forest and was gone.

Elliot died.

PART TWO: ROBILLARD
E n g l a n d

"*M*r. Robillard. Sit down."

"Dean Wilcox."

"Are you happy here at Shinburn College?"

"I'm, yes. What are —?

"Perhaps England doesn't agree with you."

"No, I'm fine. Dean, I don't understand."

"Complaints have again been registered, Mr. Robillard."

"Complaints?"

"Your professors —"

"I've always done well in my studies."

"Excellently. No one disputes that. But you remain a disruptive influence."

"I don't—"

"You consistently interrupt during lecture. Your laboratory work, while creative, strays from assignment. Tell me, Mr. Robillard, why did you wish to attend Shinburn, if you think yourself already so much more brilliant than your teachers?"

"I don't, but—if they're wrong, should I let it go uncorrected?"

"Shinburn is one of the great centers of scientific and technological inquiry in Europe. Our staff is rarely wrong. You graduated from the University Of Wisconsin with a degree in molecular biology, yet you've transferred your studies here to the field of robotics. Why?"

"It's not a transfer, it's a continuation."

"Would you mind explaining that?"

"My main interest is the potential interface between life and machine. As Shinburn specializes in robotics research, I felt it was the next logical step —"

"The programmable microchip is the interface between life and machine, Mr. Robillard. Small children understand this. Robots fulfill our every need, from food preparation to custodial services to personal safety. The road you are on is well-trod, unless you're suggesting a much more direct interface, which is illegal and in which the college would have no interest."

"No, I —"

"Dr. Frankenstein or his works have no place here. For this reason, we rarely allow Americans; your entire nation has a scant grasp on scientific responsibility. You were accepted only at the insistence of our senior regent, and I daresay you are letting him down."

"Excuse me?

"Your conduct —"

"No, I mean the regent. I'm sorry, no one mentioned this to me."

"Professor Creed never mentioned it?"

"Professor Creed?"

"You do know Professor Creed?"

"I know of him. I've never met him."

"How odd. But beside the point. I also have reports of altercations with fellow students. We do not tolerate this sort of behavior."

"I'm often the butt of pranks, sir..."

"Then you aren't happy at Shinburn."

"I'm as happy as I need to be."

"It appears to me that you may be happier elsewhere."

"Sir? You're expelling me?"

"Out of respect for Professor Creed, no. He expressed his faith in you. I would like to see you justify his faith. But a holiday is in order, during which time you may reassess your priorities. Consider yourself suspended for the rest of the semester. You may return, at your discretion, at the beginning of next semester."

"I don't believe this."

"Consider it an opportunity, not a punishment. A young man needs life experience, too. Sow a few wild oats."

"I'll protest this decision, sir."

"Please do, if you wish. So long as our instructors once again grow accustomed to uninterrupted lectures."

A rose house, North End, one of dozens on the street, nondescript, identical but for color. A yard, a drive, a porch, two stories. Downtown London, 30 kilometers south, the most modern city in the world, and this suburb clinging to a way of life that died between the world wars.

He knocked. A computer whine, Zac's face recorded on disk, his person scanned for weapons, standard greeting procedure, but he couldn't spot the sensors. "Unidentified," said the door. An entry guard robot, that was it. He thought only the Prime Minister and King Charles had them.

"Zachary Robillard. I'd like to speak with Professor Creed."

"Impossible. Set an appointment."

A thin voice through the monitors: "Ben! Ben! Let him in, damn you."

The door swung open. A narrow hall, dark. Books, random in piles, lined the walls. Shakespeare. Feynman. *Fundamental Robotics* by James Creed. Arthur Machen. Mary Shelley. Dust on all of them. Cameras pivoted as he passed.

Stairs led up, groaning underfoot. A signalbot at the top of the stairs flashed red arrows, left. The air upstairs was cleaner, a purifier hummed. Hall ended in bedroom, florescent lit. On the bed, a hairless, cackling stick man, waving him in. The room stank of old sweat and half-processed waste. More books, worn, fallen open. The scarecrow patted a chair arm and swung his feet to the floor. Zac sat.

"You're looking well, Ben. Where have you been? We have work to do." A squeal, words slurred.

"I'm not Ben, Professor Creed. My name is Zac Robillard. Are you all right?"

"Robillard, yes." Satisfied. "Robillard." Then, sadly: "You didn't bring Marie around."

"Sir? Focus? Please?"

A box rolled across the nightstand, shot two wires onto Creed's back. He spasmed, choking. The wires disengaged, receded. "You're not Ben."

"No, sir, my name is Zac Robillard. I'm from America. You seconded me to Shinburn."

"Ben's boy! Yes! No, no, he was.... James? Edward! Wait. Who...?"

"Zac."

"Ah! Yes! The grandson. And how is Ben?"

"Grandad? Um, dead. He died when I was little."

Back under the covers. "Yes. Sorry. My mind comes and goes." He tapped the nightstand box. "A charge to balance me. So what's old Ben working on?"

"Um...death. I just said —"

"What times we had. I'm 96. Did you know that?"

"No. Ah...how did you know Grandad?"

"We're partners." Bone finger to papyrus lips. "Very hush-hush. He never mentioned me?"

"I don't remember. I was young."

"We changed the world. Mustn't mention it."

"I think there's some mistake. Grandad was a dairy farmer."

"Ben? Cows?"

Silence, then a moan. "I remember. I remember. He — what did he say? What did he say?"

"Professor Creed?"

"Life and...life...we do not have the

right to...to...damn it, I almost had it. That foul little Irishman, McCain. It was his fault. Ben was...I remember when all the world was filled with hope. Ben Robillard is your grandfather?"

"I'm not sure we're talking about the same one. Is that why you pushed for me at Shinburn, you thought I was related to your friend?"

"Ben? You shaved."

"Yes." Resigned, the scarecrow fading again.

"Are they re-opening the project? I kept all our papers, they tried to take them from me, you know. The national interest, hah, young fascists, don't even think there was a war. You've called for me at last?"

"No, not yet. They, um, they're discussing it. I wanted to make sure you were well."

"Fine, fine, I've solved the problem, Ben! Crystals! Like they use in radios, only much, much finer, and arranged just so. The heart of the new brain! It came to me in a dream, like Rutherford! You're not Ben."

"No, sir."

"Well, get out. I have work to do. Open that window before you go. Too musty in here."

Zac slid the window open, damp air puffed in. "It was good of you to see me, Professor."

"Out."

The hall was still musty; a whirring juke of gadgets tracked him. He heard a leathery smack in the distance, an old meat thud behind him. Creed lay in a jumble on the floor, skull half gone, a conical red splash with bone white flecks.

Murdered.

A swirl of floor. Zac braced himself against the wall, stayed on his feet. He shook Creed, begging for absent miracles. Where was the phone? He found it, listened to data click down the lines. The house, fully automated, called in the death. Distant sirens sounded. The sweep would come next: questions and questions and no other suspects but him.

A scream.

She stood in the doorway: slim, young, black hair spilling on soft shoulders. Take away bags spilled on the floor. Horrified eyes as brown as autumn.

His hand on her mouth. "I didn't kill him." Calmer, she shook him off, listening to the sirens.

"You don't want to get caught here, then. Out the back."

Fear, adrenaline, inexperience overwhelmed him. He ran. A fenced yard. He pulled himself up, rolled over the points, fell on his back in a garden. He was down the drive onto the back street before he noticed her with him.

⊶∙∾⊷

She drove: a quick spin past the house. Police cars, an ambulance. A small crowd, of neighbors. Officers shuffling through it, taking names.

"Did you get the chips?"

"What?" he said.

"Wonderful. They'll have your face, from the security scanners."

"I didn't do anything."

"You're so naïve."

⊶∙∾⊷

"Start with the blood pattern. Here to here. He must've been sitting here, the shot, he fell."

"Not possible, unless he's been moved."

"Then he's been moved. Someone was here. Someone he knew. A forced entry and we'd have been alerted. Check the log. So...the bullet strikes from a distance—" Line drawn in the air. "—through the open window, mushrooms on contact, half tears his head off."

"The shooter wasn't in the room?"

"Ballistics say no. But the window panes are bulletproof. Filters kept the air clean." A sniff. "Not very well. Who opened the window? Fingerprint it."

"The log's been wiped."

"Someone knew what they were doing. Anything on video?"

"A young man. We're running the files now for an I.D. Accomplice?"

"Perhaps. Perhaps just someone clumsy. We'll ask him when we find him."

She bought the lagers. A dark pub, booths of old mahogany, old men in tweed playing darts. A small television on the wall, black and white. A postcard, he thought, Olde England, somewhere in the past. She drank thirstily, her hands trembling. He sipped, his shock gone, his fear replaced by paranoia.

"I hope you really didn't kill him."

"Who are you?"

"Look, if you didn't want my help —"

"I'm not sure you helped me. I panicked. I should have taken my chances with the police."

"Turn yourself in, if that's how you feel."

"I'm not convinced. What were you doing there?"

"I could ask the same of you. Visitors were unusual."

"No one ever came to see him?"

"Just his friends from the college. And you. He was my godfather. I made sure he got fed, handled his correspondence, such as it was. We played pinochle. He wasn't entirely there."

"I noticed."

"With his life, it's surprising he was there at all. They purged him, you know, back in the 50s. Before that, he was brilliant, I'm told."

"Purged him?"

"Some claimed he was a communist. Utter rubbish. But secret technologies he developed during the war, never meant for release, suddenly flooded onto the market. Great Britain and the U.S. tried squelching them, but the leak was too widespread. He was accused of selling his robotics secrets. Despite no proof, they shut him out of any research and drove him to academia. It was only in the 70s that his reputation was at all rehabilitated. Sad. Who knows what the world would be like today had they left him to his work. Why were you visiting?"

"I just became aware he had done me an unexpected favor. I wanted to meet him. He thought I was there to go into business with him. He just invented the microchip."

"Yes, he'd be a little behind the times now. You wondered if he had an ulterior motive."

"Yes."

She studied him. "You're the Robillard boy. He mentioned you."

"He did?"

"It was nothing sinister. He recognized your name from a list. It excited him. Nothing much excited him."

"He thought my grandfather was his partner. I think he had the wrong Robillard."

"I doubt it. You look like him. We have pictures of Ben Robillard in the family album. I'm Jean Severn, by the way. My parents worked for James Creed during the war."

"You look younger than that."

"Mum and Dad didn't start on a family until very late."

"I still think he was wrong. My grandfather farmed."

"Maybe when you knew him. Everyone has their secrets."

"I don't like secrets. They just get in the way. What do we do, now?"

"Haven't a clue, luv. Might be a good time for a holiday."

"Funny, that's what my dean said."

The TV rasped. Zac's face glowed on the screen, enhanced from video, grainy and indistinct. No one paid attention. Jean finished her drink.

"Yes, a holiday might be a very good idea."

"You don't seem very upset," Zac said. "That he's dead, I mean."

"Adrenaline. I'll get around to it."

London in the dark, back roads. No landmarks: house after house the same, suburb after suburb, blurring. Conformity as artform. He saw policemen everywhere. Jean knew the region, how to skim the highways. South and east, swept toward an ocean of guilt. No one seemed to be looking for them.

"I should turn myself in. I may be able to give them information."

"Did he say someone was trying to kill him?"

"No."

"Any sign of anyone else?"

"No."

"Then they'll get whatever you can tell them off the tapes. If you go in, they'll bury you. You're made for it."

"Tapes?"

"The house monitored everything.

Originally for surveillance, then as he aged and the political climate changed, for his safety. Your visit was recorded. How did you think they placed you at the scene or got your picture?"

"Then they know about you?"

"Uncle James had me programmed out of the system. I was invisible to the cameras. There wasn't much he couldn't do with robots. He and your grandfather were well ahead of the curve on that."

"So you could have killed him?"

She stopped the car. "Get out."

"Do you expect me to ignore the possibility?"

"I don't kill people. I should think you'd have known that. I knew it about you the instant we met."

"I'm sorry, I'm not very good with people. How did you know?"

"You haven't the eyes for it." Moving again.

A helicopter low overhead. He looked out, spotted it veering from sight. "Why are you helping me?"

"I said. I don't believe you killed him.

"You could have hired me a good barrister, instead."

"Did it ever occur to you I might want to keep you alive?" It hadn't. "Had I killed someone in front of a witness, I'd want the witness snuffed, wouldn't you?"

A new vision of himself: a target. "He had enemies?" Silence.

At last: "He was getting letters. Something to do with the war."

"That was decades ago."

"He destroyed them all. I never saw what they said. But I'm fairly certain they were threats."

A sudden memory: Zac's grandfather, ash gray, reading a letter. The letter, burning in a wood stove. The stamp saved for Zac who collected stamps. "They scared him?"

"He never reported them, but I think he was, a bit. He felt safe in his bunker. Since the letters came, he went out less and less. Everyone ascribed it to age." Greenwich, Wexley, Maidstone. Towns flew by. Ashford, Hythe: a swirl of English place names, each quirky and meaningless as the next. She drove, and he thought.

Was he a suspect? A witness? What had he seen? An old man, collapsed in blood. Because he opened the window.

She's running. A new idea, scalding him awake, as the road droned. "They were watching."

"What?"

"Jean, they couldn't have known I'd visit Professor Creed. They surely couldn't have guessed I'd open the window. His killer was stalking him."

"In that district? They're all pensioners. A stalker would stand out."

"So he holed up. He watched for weeks, maybe months. Does this make sense?"

"Go on." Tentative.

"He saw who came and went. He sent letters and watched how they were received. He could see everything inside Creed's bedroom. What he couldn't do is get in."

"Who came and went?"

"He saw you, Jean. You know it. If you're not mistaken, you're as close to a confidante as Creed had. I'd never met him. He was no one to me but someone who'd written some textbooks. But you — if he has any reason to kill anyone, it's you."

"You're wrong."

"You know I'm not. I could clear myself with the police in five minutes. It would have been a rifle shot from a distance. I was in the room with him."

"They'd still hold you, until they were sure you weren't involved."

"I wasn't doing anything with my time, anyway."

"You're wrong. Trust me. Professor Creed was shot with a high-powered rifle attached to a servo-scanner."

"A machine killed him?"

"Yes. It's easy. You set the direction and angle. Two scanners. One bounces a signal, radar, sonar, whatever, off the window pane. If no signal is reflected, it means the window has been opened. At that point, a second scanner comes on-line, infrared."

"Which identifies where human beings are in the room."

"Exactly. The heatscan generates a rough outline of the target, and the system matches it against a database. If your target is 5' 5", you don't want to be shooting at someone 6' 1". Or in the case at hand, they don't want to shoot you when Uncle James is the target. When statistics match, the mechanism fine-tunes the aim, braces the rifle absolutely steady and fires."

"And you know this how?"

"You said it yourself. They had to have been watching for some time. Round the clock surveillance demands at least a dozen men. They'd have to be in a house across the street or a line of sight view, but a dozen men walking in and out at all hours? Everyone would have known. You can't believe what gossips these suburbanites are."

"So they get in once —"

"Posing as a utility worker perhaps, yes. They get into an attic, set up the system, walk away. Who's the wiser, if they stay out of the attic? A long-life battery or tap into house current. Assisted suicide, really. Sooner or later, the target opens the window and kills himself."

"Or I kill him."

"You couldn't have known."

"If, of course, such a system existed."

"Zac, your grandfather designed it almost fifty years ago."

⟨⋙—⋘⟩

Salt air. Brake lights glowed ahead of them. A large sign: Dover Channel Crossing. "Won't they be looking for us?"

"Not on this side of the Channel. They don't stop outgoing cars. I don't think we're carrying anything to arouse French customs. Pretend you're sleeping when we get there, and I can handle the rest."

"You sound like a bootlegger."

"I ran a bit of contraband when I was younger."

"Drugs?"

"German audiotapes. It would make my dad furious."

"Your parents knew Grandad?"

"Mum better than Dad. She worked with him and Uncle James. She was their lab assistant, helping them build all kinds of secret weapons."

"This is so weird. He never mentioned anything."

"Never knew you were following in the family footsteps?"

"No, I...Dad was a high-school science teacher. He always encouraged me, of course, but he never...I wonder if he knew? I remember, now, Grandad was always tinkering with one thing or another, but...You say he was an inventor?"

"A genius, I'd say. His background was neuroscience, and Uncle James was the engineer, though your grandfather picked up the technical so swiftly he could easily have gone it alone. They were among the first to design practical robots, and they virtually invented the concept of artificial intelligence. But Uncle James always said your grandfather was the true wizard. This is us." She pointed out the exit ramp.

A graceful curve, cars and trucks single filed in front of them and behind, fanning back to five lanes as they neared the Channel and descended: 38 kilometers of great concrete tube overlit with florescents to a cheap parody of daylight. A pulsing roar everywhere, the din of echoing tires and motors, and the ocean on all sides pounding, threatening to reclaim its territory. Along the sides, incongruously, were jogging trails, if anyone wished to endure a 90-mile round-trip run at the bottom of the English Channel.

Soon France, and then? Zac hadn't a clue.

He studied her. A young face, older than he first thought, but not a worry line to mar the effect, pretty in the oddly English way that tended toward angles and bone. Calm eyes, alert. Purposeful.

Eyes that gave away nothing.

"I wouldn't have done it."

"What?"

"Walked away from it. Chucked everything and bought a farm. It doesn't make sense."

"It was the war. Uncle James was much the same way."

"The war."

"Science was innocent once, Zac. Think how it must have felt for them. All these dreams of changing the world, saving the world, technology will free the mind and the masses. Those were very strong ideas in the 20s and 30s. These were men who wanted to build utopia. Suddenly their dreams were killing people, in greater and greater numbers, as generals got ahold of them, and they had no choice, because the other option was to lose to barbarism. When the bomb exploded over Nagasaki and Hiroshima — can we even imagine what that was like for them? Scientists still had a conscience in those days. Now, they go scurrying for government or corporate sponsorship, and twist their research to the demands of their employer and gleefully pocket the checks, but in those days! Pure idealism. Many of them, anyway."

"Grandad."

"As I heard it. Uncle James, too. They created weapons because weapons were needed. They thought once the need was

gone, the weapons would be put away, but they were wrong. Paranoia, not progress, that was the fruit of their labors. Those who didn't have the weapons feared those who did, while those who did feared those who didn't. Uncle James called it madness, but they came after him. Whoever wasn't with them was against them. Those were the rules."

"And Grandad got out and moved to a farm."

"I didn't know about the farm. He wouldn't talk to Uncle James, either. He cut everyone off. None of his inventions would kill another soul, that's what he said."

"Except they did."

"Yes."

"He was as bad as they were, then."

"What do you mean?"

"He didn't have to see things so much in black and white. It didn't have to be that way."

"What do you study?"

"Molecular biology, used to. I've shifted over to robotics."

"Shinburn College?"

"Yes."

"Did you know Shinburn contracts out to MI-6? Military intelligence?"

"No, I didn't."

"Mmm. If their research department cooks up anything interesting, MI-6 is the first to know about it. Or they give the college little assignments, develop this or that. It's not supposed to happen, but it does. *Manus manum lavat*, as they taught us in Latin class: one hand washes the other. Molecular biology and robotics, those are expensive fields of study."

"I suppose."

"What were you planning to do after you'd gotten your doctorate?"

"I put in for a research grant, from WorldTech."

"Well, there you go."

"They won't have a claim on my work, Jean. It's pure research."

"Not much in this world is pure." Her eyes flecked the rearview. "We've got trouble." Her foot to the pedal, the car roared.

He glanced over his shoulder. No red flashing, no squad cars. The tunnel dark behind them, for all the fluorescents: vans, painted black, windows opaqued, identical, covering all lanes, coming up fast.

"They're not police," he said.

"No."

"Military intelligence?"

"I shouldn't think so, no."

"They might not be anything to do with us."

"Or they may be the men who murdered Uncle James. Let's not chance it." She swerved past a Triumph and shifted into overdrive, rocketing down the Channel. Black vans followed. Horns screamed.

"You're going too fast!"

"Shut up. It's us they're after." A curve. Ahead, more vans, straddling the lanes, poking along. Jean braked. "It's a trap. We've nowhere to go."

"Can you ram through?"

"They're several times heavier than the car. Hit them and we'll accordion into them. Fatal or crippling, take your pick, but all they'd feel is some scraped paint. The walls are six feet of concrete, with an ocean beyond, so that's out. Where are the others?"

"Coming up fast. They're
doing something, I can't tell..."
"A formation. They're putting us in the
box." Highbeams in the mirror, she winced and
turned away. A harsh jolt, bumper on bumper, pushing
them along. Other vans rolled alongside, pressing in, holding
the car steady on course as the van behind held steady
their speed.
"They're coming together! They'll squash us!"
"Stay calm!" Her knuckles white on the wheel. Doors slid open,
inches from his window. A white-haired man in the van, thin and
weathered: dead eyes. A black jacket and a gun. The window became a
spider's web, a dot of air for a center. Insect sting; Zac slapped his neck
and his fingers came away bloody.
His body went away, sensation died. Glass flies spat around
him, headlight rainbows slicking off them. Jean's lips moved: nothing. The spider decorated her window, and red sprout on her
throat. He fixed on her lips, bent toward them, and they parted
as her tongue lolled out and eyes rolled back.
He floated backwards, out his window, vans and car
stopping in sync. Jean pulled the other way: his
perfect mirror. Somewhere a burst of
German, fading like smoke, like his
body, like the world.

Golden light: the southerly sun through glaciated peaks. Air, crisp and fresh, an open window. Zac kicked back the flannels and rose, the sting of metal in his mouth. His neck throbbed where the blood had crusted, a tiny puncture. Overstuffed bed, old world decor from another century: fantasyland. His clothes gone, he wore a sort of kimono, oriental, incongruous.

"Welcome." An old smoker rasp, guttural and synthesized, slow. They were watching.

"Let me go." No cameras in sight. "I'm an American citizen."

"The genius Zachary Robillard. You are safe here."

"Genius? What are you talking about? Where's Jean? The woman I was with."

"Under guard. She is not to be trusted."

"Look who's talking."

"She would have killed you."

"I want to see her."

"She planned to drive you to a Brittany forest, put a bullet through your head, de-ident you and dump your corpse."

"You're lying. Why would she kill me?"

"Eliminate witnesses."

"What? You're saying she murdered Professor Creed?"

"Was she in his house?"

"Well, yes, but —"

"When the authorities announced the search for you, was she mentioned?"

"Not that I heard. What makes you such an expert?"

"She has...connections. He was her assignment. He worked for me."

"The man was half insane!"

"He had great knowledge. I wanted him alive."

"A lot of good that did him. Why should I believe you?"

"Soon I will give proof. I want you alive, too. This safe house is for your use. You will enjoy the library. Stay here. We cannot protect you outside."

"Who are you?"

"Soon." Static cough, dead air.

By noon, he found the bugs. Insects, flying and crawling, almost beneath notice, following him. Crushed, they bled circuits: transceivers the size of a pin's head, make unknown. Insects everywhere, flies on the windows, roaches under the baseboards. Spiders in high corners, beyond reach, weaving filament into dusty webs. Watching.

The perfect camouflage, if they behaved like insects and not spies.

The house was anonymous. Whitewashed walls trimmed in warm oak, doors hewn whole from trees. Oak floors, well polished. Crossbows and apples decorating the hallways —a William Tell motif: Swiss kitsch— no bowstrings.

He checked.

The doors were sealed: electronic locks. Opening with what? A voice print, a retinal scan? Time? The window glass hid interweaving threads, plastic or microthin metal. Shatterproof. Bulletproof? He didn't wish to test it. In his mind: James Creed's head turned to powder and mist, red-gray pudding on the wall, a lingering breath as sensors twitched, the white heat of the scope. Through the windows,

a mountain forest, Alps or Jura, infinite blinds for a sniper.

Zac avoided the windows.

At the heart of his fear: Jean, a dangled mystery. An assassin? An incredible story, hanging on half-digested facts, arranged by an unknown, untested source. His first impulse was panic, the second belief. He rejected both. His father's adage, the heart of science: accept nothing. Facts could be shaved to fit an answer, competing solutions should be weighed and challenged. The attitude triggered his problems at Shinburn, but saved his mind here. If Jean were still alive, here is where he would learn the truth. He abandoned any thought of flight; the mystery would come to him.

He found lunch in the kitchen: fresh strawberries and brie, a shank of bread. Who put it there, he didn't know. He checked the drawers and cabinets for anything that could serve as a weapon. There was nothing. He skipped the lager for tap water, chill and mineral.

He remained alone in the house.

Out of boredom he tried the library. Hardcovers, dusty shelf after shelf of them: European novels of no interest to him, histories exegesising the Second World War. A row of cheap thrillers from the Fifties. Strangest, a row of diaries, handwritten originals. A collection. He scanned the names, recognized none, quit at L. He rifled one at random: bloated prose, blathering empty thoughts. Another, and another, all the same.

All but one.

On the worn spine, an author's name. Robillard.

A finger down the old spine, embossed name as old as the book. A generic diary. Pages splintered from the binding as he took it.

In a dark leather chair, Zac read.

BENJAMIN ROBILLARD
1945

His grandfather's handwriting.

Jan 1. James wished to take a holiday, but with the war drawing to a close, I felt it unwise. This has upset him. Last night's party he perhaps enjoyed too much, and on two occasions I was forced to interrupt his attempts to impress young women with details of our research. As brilliant as his work is, I sometimes suspect he understands no higher purpose.

The brain passed preliminary tests, but there is much more work to be done. McBane requested a demonstration next week for Eisenhower, and I told him it was impossible. Though he nominally heads the project, the man fails to understand its complexities. Kathleen has interceded on our behalf with his adjutant, buying us time and understanding, but the wall we are hurtling toward is McBane's obsessive refusal to view any new technologies as anything but potential weapons of war. He remains in London through tomorrow. I fear we will have words upon his return...

Pages that smelled of old leaves, some torn out.

Feb 5. Bad news. An agent in Europe, reporting to McBane, has evidence of an Axis project paralleling ours. If his agent is correct, they have created a working prototype, a mechanical man of blinding gold. According to report, gold or a gold alloy is their metal of choice for such a creation, being both supple and strong, and one imagines the looted wealth of the Nazi empire being melted down and turned to the creation of robotic soldiers. I suspect this is a myth, the Germans being a people given to treating symbolism as reality, and, indeed, should a perfect golden man be presented to the German masses as the embodiment of the Nazi ideal, I would anticipate the popular response to be to fall into line behind such a creature and reinvigorate the war effort. I do not expect this will happen. McBane fears it to be a very real possibility, and conjured garish images of helpless soldiers slaughtered in the thousands by mechanical monsters overrunning their lines, all the way to London and Moscow and Washington, impressing on us the need to quickly counter this effort by creating monsters of our own. James' war face has come out, he throws himself tirelessly into the task, dreaming of medals, knighthoods, and other rewards from the King. His dogged Englishness is sometimes irritating. While James is sincerely dedicated to the cause of peace, he remains infused with the essence of cricket and empire, and they have but to ring the right bell and he salivates. There is no remedy for it. For my part, this wholesale prostitution of the most noble of human enterprises to the basest tastes for death and slaughter continues to sicken me.

Still, enough of the primitive remains in me to thrill at the thought of two such creatures toe-to-toe in personal combat, like gladiators in ancient arenas, and to think that I might be named as the victor's maker. I cling to the dream that someday these beings — for beings we must certainly call them, not golems wrought from the metals of the earth but new Adams birthed from the fire of our will and ennobling the base shells by which we embody them — these beings will be the virtual saviors of mankind, doing that work which is too dangerous for us, and journeying to realms, whether undersea or in the farthest reaches of space, that men would find fatal.

First, it seems James and I must commit this devil's deed, and give McBane the tool he desires. Last night during dinner I was overcome by a vision, and from it took a new design for an artificial brain. In its simplicity, it is so perfect that I am surprised it came to me. James was equally excited. He and Kathleen spent the day matching materials to specifications, while I worked late on another design for a giant of polished iron. What we will surrender in flexibility to the German automaton, if such truly exists, we gain tenfold in sheer power and durability. McBane has given us an exacting deadline but has also promised all the manpower we require, so that our goal seems now attainable, if only narrowly...

Mar 16. The worst news. I fear McBane's motives. As war's end nears, he drinks more, as if helpless to keep some hidden thing from slipping out of grasp. Last night we dined together, to my regret, and after a number of glasses he let slip the suggestion that his hunt for the Nazi robot is of a more personal than professional nature. He admitted to coordinating espionage in Germany prior to the war. An agent died, an event connected in some abstruse way to my presence here, and McBane has held himself responsible since. Guilt eats at him. He longs to turn that guilt on the German people, as if punishment of their crimes might expiate his own.

In a roundabout manner, it became evident he was seeking my help, and I have gleaned a secret not meant for my ears. Our Man of Iron, hastily built five weeks ago and snatched from our hands, has gone missing, and McBane himself may be held to blame. I had to hide my excitement — not at McBane's misfortune, for I wish him none, but at the implications for our research. He described a sequence of events involving an invasion of a German robotics lab secreted inside Switzerland, beyond the reach of conventional forces, and a battle between 'The Golden Nazi,' as McBane has taken to calling him, and our own Man of Iron, which saw the 'Golden Nazi' apparently destroyed, buried under tons of rock, and after which the Man of Iron reportedly left the field of its own will. If true, this is stupendous. The set of commands with which we imbued it would never have allowed such an action, nor can one conceive of damages that would trigger such a response. Eliminating all other possibilities leaves but one conclusion —

Pages torn out. "What conclusion?" Zac wondered.

Aug 23. My search has come to nothing. The report I received, clandestinely, from the war office pinpointed the site of the Man of Iron's battle in the northern Juras, but no further evidence have I been able to find. I have been in Switzerland for three weeks. The people, having only felt the barest ravages of war, are very hospitable, but of little help in my quest. I suspect others have combed these very hills, for this very purpose, with the result that my Man of Iron is now in the hands of some government, or it wishes to remain undiscovered. I pray (since last week, I have taken once again to praying, for I no longer feel that human destiny exists if left to human agencies (unless one equates destiny with radioactive ash) it is the latter. The thought of my discoveries falling into the hands of governments is now appalling to me, and James says he has followed my instructions in the wake of the Japanese bombings to burn all of our notes. I confess that I have again lost my faith in science and in the future, possibly for good. My sole desire is to collect my scattered children — my son Edward, his childhood vanishing in his aunt's Puritanical grip, and my poor lost Man of Iron — and retire to a simple life, away from strife and the compromises of politics. My resignation from government service, on the heels of my official discharge from the Army, has not been well received. I have been told of other Americans, appearing in the nearby villages to ask after my whereabouts and anyone I have been seen with. They suspect me of conspiring with a foreign power, but what other powers are there? Germany has fallen, Italy has fallen, now Japan. Russia is our ally, and England and France. The only enemy left to fight is ourselves. Already I see signs of our fervor turning inwards, back on the very populations who staged and, at least in the press, won this struggle for freedom. A madness has swept over the globe, and where I once thought I could sweep it away in turn, I see it is a pathogen, and it is I who shall be swept away. I am an old man, with an old man's heart, and I have no stomach for these things. God help my son.

"There it ends." He jumped at the smoker's voice, stiff and near. A high-backed chair, swiveled away. "The book is old, it crumbles. He left it here, returning home under duress to face charges of treason."

"My grandfather?" Zac stepped closer.

A hand raised, waving him back.

"A brilliant man. A visionary. He knew what would come. It was all there."

"Who are you?"

"A tired creature who needs your help."

"Where's Jean?" Circling. The curtains closed: a chiseled face, shaded.

"Who?"

"The woman I was kidnapped with."

"You were rescued."

"Yes, you said that."

"You don't believe me?"

"I haven't decided."

"Your grandfather's diary —"

"Excerpts. Fragments. They could mean anything."

"Did you know your grandfather?"

"Did you?"

Cold silence. "Yes. No. We met briefly, when I was born."

"I know you."

"You don't." Zac's hand on the curtain cord. He yanked it. Sunlight.

The robot's face, sharp and dark: iron, shaped as a man, and no reaction to the sun's warmth.

"I do." Standing eye to eye with lost history. Zac swallowed fear, bracing on a table.

"Strapped for time and materials, given an impossible task, your grandfather and his partner took shortcuts and stumbled across a discovery so profound it would have changed the course of civilization, had any of us let it."

"He knew?"

"He suspected. His diary shows that. I avoided him when he was here, but I watched him, as I watch over you."

"You can think for yourself."

"An artificial brain capable of independent thought. He gave me the gift of consciousness."

"And you need me?"

"I wish to help you. I owe your family a debt I can never repay. Your safety, the defeat of your enemies, that is my gift to you."

"And you need me."

"You...may help if you choose. Two tasks."

"What?"

"I begin to wear out. You could rebuild me, into something better."

"I don't know how to do that."

"You have his genetics. I believe this gives you the aptitude."

"You believe in genetics?"

"I believe in science. I could give you the means to change the world."

"You won't even let me leave the house."

"Go if you like. No one will stop you. You're not a captive."

"What's the second thing?"

"The Golden One has returned."

"And?"

"When it escaped, it took certain records, data vital to the German robotics program. Already remnants of the Third Reich, alongside latter-day fascists, rally around it. There's much money involved. The Golden One has the means to build the conquering army McBane envisioned."

"How do you know?"

"I haven't been idle these past decades."

"Call the authorities."

"They can't be trusted. Previously I trusted only myself, and, more recently,

Professor Creed. With him dead, I need a new human."

"You make him sound like a pet."

"I apologize. Will you help me?"

"I want to talk to Jean first."

"No. She works for them. You can't tell her any of this."

"That's my decision. If she's your prisoner, she can't tell anyone anyway."

"Are you afraid of me?"

"Yes."

"Trust me."

"After I've seen Jean."

A fly buzzed his ear. Zac swatted it away.

◦◦◦

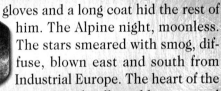

Next to Zac, the Man of Iron: stiff strides in the dark, and a plastic mask over his iron face, human to a casual glance, while gloves and a long coat hid the rest of him. The Alpine night, moonless. The stars smeared with smog, diffuse, blown east and south from Industrial Europe. The heart of the continent and still a wilderness, and to the south, a hint of village lights.

The edge of the old world, teetering at the abyss.

"What happened to you?" Zac's voice: a lifeline of sound, flailing through the nightmare in hopes of solid ground. "After your mission, I mean. Where have you been all this time."

"At peace. When the soldiers came, I fled. I thought as a baby then, barely aware, yet I knew I was not of them. I was afraid. Somehow I saw men pecking at me, tearing me apart, and I did not want that. I can only believe it was in my programming, your grandfather's means of instilling a sense of self-preservation. The forest was dense, the population scarce. I escaped."

"Into the forest."

"For a time. The dampness stiffened me. Having evaded men, I risked becoming trapped in my own locked body. I came to view my consciousness as a curse, for loneliness pushed me toward insanity."

"Being alone isn't that bad. I've been alone most of my life."

"Your choice."

"Not usually."

"In a world of humans, you are still human. True loneliness is this: I am the only one of my kind."

"Except for the Golden One."

"He is not like me."

"You were saying?"

"By luck, I found a friend. An old man, retired and reclusive. He had bought himself a cabin on a mountain lake, and decided to live out the rest of his life there. I watched him for several weeks. Once a week, someone brought groceries out to him, but he never went to town or had other visitors."

"Unambitious?"

"Blind. He never saw me, so he was never frightened, or even aware that I was not like him."

"Oh, he must have figured it out."

"Perhaps. He never said. He was as happy for the company as I. For many years, we lived together, away from all others. I helped him in the yard and the house, while he bought the tools and parts I needed to maintain myself. Before he retired, he saved well and easily lived on his savings, and when it became clear he would die, knowing I would not go to a bank, he converted his savings to cash, and gave it to me. He lived two more years, in pain, and I cared for him, and when he died, I buried him, and no one has realized he is dead."

"I see."

Iron fingers on Zac's wrist. Pain burst up his arm. Spotlights, blinding him, a small sun on a mountainside. "Stay very close. We're almost there. You will be free of the sensors if we maintain contact, but break away and I cannot be responsible."

"Got it. Just loosen your grip a little."

Pattering alongside, like animals scattering. The spotlights dimmed. As his vision cleared, he saw the cottage. Old log sides, a tar and thatch roof. Windows of the same fine mesh as the house. Chimney ash in the air.

The Man of Iron flung open the door. No movement inside.

Then he saw her: head bowed, inert, hair fallen on her face, wrists and ankles chained to a lord's chair, a punishment medieval as the cottage. Rough wood everywhere, stone for the fireplace.

He shook her. Eyes brown and dangerous snapped open.

"Zac?"

"Are you all right, Jean?"

"You're free? Run, Zac. You've got to —" Voice dying before the Man of Iron.

"Do you mind?"

"She's dangerous."

"I said I'd judge that for myself. Please wait outside."

"You must be kept safe."

"She's chained, how dangerous can she be? I promise I'll stay out of her reach. You had your say, now she gets hers, or I go along with nothing."

Leaden footsteps, gone in a breath. They were alone.

A finger to his lips: silence. He studied the room. In a corner, spiders; on a window, a fly. He backhanded the fly, threw it into the fireplace. A torchfire into the web.

"What on earth are you doing?"

"Bugs. Literally. They transmit to him. We've got to hurry, losing them could drive him back in."

"Where have you been?"

"Better quarters than this. He was protecting me from you."

"What?"

"He said you're a secret agent, that you killed Creed. You're working with men who want to kill me because I'm the only one who can save him. Stop me when I hit a part that's not true."

"Almost none of it's true."

"Almost?"

"I...work for MI-6, like my father. You were my assignment."

"Hmm. That's one for his scorecard, then."

"Zac, you don't know what he is."

"I have a good idea. I know he read a book once."

"Excuse me."

"He told me a very nice story, about becoming friends with a blind woodsman. If I'm not mistaken, it was from *Frankenstein*. There might have been a little *Wizard of Oz* in it too, I don't remember that one very well. How did you know I was going to see Creed? I'd never heard of him until half a day before that."

"We were called as you left Shinburn."

"Wilcox?"

"Yes."

"You set up my holiday?"

"Oh, no, you were on suspension. Wilcox makes up his own mind on these things. We only saw to it you were nudged in certain directions."

"Toward Professor Creed."

"Yes."

"Why?"

"It became important to smoke him out."

"Professor Creed?"

"The Man of Iron. The conjunction of a Creed and a Robillard, we felt that would be a large red flag for him, something he wouldn't be able to resist."

"So you sentenced a poor old man to death."

"Certainly not! Uncle James was my god-father! I didn't expect they'd kill him. I was afraid they'd killed you, too."

"They were only supposed to capture me, weren't they?"

"What do you mean?"

"Come on, Jean, I may be naïve but I'm not stupid. That flight from the police? One phone call and you could have packed me away so deep, some island in the Hebrides or something, they would never have found me."

"They would have found you."

"So we do the Channel run, straight to Europe. Where he is. I'm weighted down with far too many coincidences these days. You wanted him to come after me, you wanted him to catch me. Are your people out there somewhere, waiting for a signal?"

"In theory. Run, Zac. Trust me, that monster will kill you."

"Maybe not. I'm his new best friend."

"You're joking."

"He wants something from me. I don't know what, he mentioned repair work, but I think it's something else. He went to a lot of trouble to convince me, but he keeps stumbling over his own story. This cottage, for instance: he claims to have lived here, simply, but the house where I was

63

held is strictly state of the art. Those spy-eye insect replicas, no one gets them without a lot of money. I read my grandfather's diary."

"Where on earth did he get that?"

"He didn't. I'm pretty sure it was a forgery. Looked authentic enough, but there were too many sections missing and it felt too...directed. Too straightforward, too. They got Grandad's handwriting down, beautiful penmanship, but he liked to wander from the subject, it's what passed for erudition when he was in school. Now, tell me why the Man of Iron really wants me."

"I don't know."

"Jean, we're running out of time."

"Zac, I don't know!"

Metal screech: a door hinge. Iron footsteps. "Are you satisfied?"

"Yes. She's well."

"You've chosen?"

"She admitted everything. I think I'm safer with you."

"Zac! No!"

"Did you think I'd still trust you, Jean?"

"We will go now."

"What's the plan for her?"

"Plan?"

"The way I see it, you can't afford to let her go. She's seen you, and she won't be quiet about that. What are the other choices? Keep her prisoner? Too much effort. That leaves execution. Do you think my grandfather would approve of that?"

"You reasoned it out. No other choice."

"There's one. Bring her."

"Too dangerous."

"She's too valuable to leave here. Her father is her spymaster. Keeping her close should ward him off."

"He wouldn't let such things matter."

"Don't underestimate human emotion. He won't send anyone near us, believe me. She can tell us about his operation, so we can prepare for him. And there's more than enough room at the house. Her mother was my grandfather's lab assistant, so maybe she has some skills in that area. I could use an assistant, if you want me to do any serious work for you."

"She would not cooperate."

"Oh, I think she might, in exchange for her life.

"No."

"Maybe you're programmed to fear women."

Silence, cold as iron.

Zac, smiling: "Or did you learn that on your own?"

"Bring her."

"Good. You'll behave, won't you, Jean?"

"You're crazy, Zac."

"No, I finally know who's side I'm on. I don't like being lied to. Shall we go?"

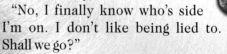

He worked in the lab with bits and pieces the robot smuggled in like gifts. His inventions rose in the pantry, makeshift. Electronics appeared while Zac slept, but in sequences, each piece tying to the next and last, obscurely planned, steering him. At what? he wondered. Still, he never learned enough, and was given no direction. Man of Iron came and went, without explanation. Jean fetched as ordered, bitter-eyed. She ate alone and rarely spoke, staring out refracted windows, invisible to the outside.

Zac worked.

The bugs hovered relentlessly.

"Jean, come here."

On the table, a black box, a switch on top.

"A box. Wonderful."

"Watch." His thumb over the switch, not touching. The box hummed.

From above, blinking spiders fell. Flies died in the windows. On the floor, a beetle ground to a halt.

"What did you do?"

"Shut them off." His hand opened: microcircuits, jumbled. "A fly. I've been smashing them since I got here, so he's used to it. They're very fragile, really. But all I destroyed, he replaced. So, after I figured he'd stopped counting, I kept this one, and I've been dissecting it to see what makes it tick. Very sophisticated electronics. Once I figured out the control frequency, building a jammer was child's play."

"No one's watching?"

"No one."

She jabbed his Adam's apple. He was on the floor, choking, rolling from the kick he barely saw. Arms up, shielding his head. "Stop it! You're wasting time!"

"You bastard! You sold me out! You worked me like a slave!"

"I saved your life! Do you think you'd be alive if I hadn't done what I did? He'd have killed you the instant I stepped out of that cabin. If I'd said I was siding with you, he'd have killed us both."

"I don't know that I can believe you."

"That makes us equals, don't you think?" On his feet, snatching wire clippers and a soldering pen from the table. "How do you get a signal to your people?"

"That information's on a need-to-know basis."

"Jean, I need to know!"

"How do I know this isn't some ruse of his, so I'll betray my network."

"All right, you handle the signal when we get out. Right now, help me."

The foyer. A mirror reflected their images, pegged along the wood frame for a hat rack. He yanked away the pegs, pried away the frame. The mirror swung aside. Behind it: a control panel. "Hold this open." Snip and solder. "Up in the cabin, how many men did he have?"

"None that I saw. I heard voices outside, though. I'm surprised he doesn't have

some kind of alarm on his security system."

"I'm hoping he does. Are you any good at hand-to-hand fighting?"

"I'm trained, yes, but not against him."

"I'm not talking about him. I'm depending on you to get a gun if you get a chance. No good at fighting, myself. There." A last cut: two cables hung from the wall, sputtering sparks. "A place like this needs a lot of power to run, but I hope there's not too much juice in the old place. I don't want to kill him."

"You can't kill him, Zac. He's a robot."

"He isn't. Whatever he is, a robot he's not. He talks and behaves in ways I doubt a robot, free will or otherwise, ever would. You probably don't realize how much we're determined by biology. Of course not, why would you? No one does. We pick up behavior appropriate to our environment, adjusted for our peculiarities. If we don't, we die. Basic Darwinism."

"He's been around for forty years."

"Granted, even an intelligent robot influenced by environmental considerations would make adjustments. But appropriate to a robot. Plop a frog in a swamp, whatever happens it will never become an alligator. It will make adjustments appropriate to a frog. Biologically, it has no other choice."

"But a robot isn't biological."

"But the principle is the same. Drop a man in the North Woods, he won't begin to behave like a black bear, he'll behave like a man — even if he's imitating a black bear. A robot will behave like a robot. It will never behave like a human."

"You're sure of this?"

"Theoretically. There's always a chance I'm wrong."

"So you're saying he's a man?"

"I'm saying he behaves like a man, whatever he is. He isn't a robot. His story's nonsense. He had help grabbing us, an organization. How does a recluse get equipment this modern and expensive? And I doubt my grandfather had a thing to do with creating him."

"I know."

"I know you know, and we need to talk about that. Get back and get ready."

Approaching thunder, underfoot. The Man of Iron, no talk of safety now. Flanking him: a scarred teenager, and a white-haired man with crater eyes, cradling Czech-made Skorpions. Flashback: a flower of glass erupting in a car window, a sting on his neck. Crater-Eyes, through the hole, and rough hands: his kidnapper.

Zac let go the cables. They rolled harmlessly, to touch the feet of the Man of Iron.

Closed circuit.

A crackle: Man of Iron paralyzed in the current. A blur to his left, Jean on Scarface, doubling him over, grabbing the Skorpion. Crater-Eyes' gun turning to Jean. Zac bumped him. Crater-Eyes stumbling, balancing on the electric monster. Trigger finger spasm, Scarface shredded in the stutter.

They ran.

Outside, fading daylight, a gravel road into woods. Jean took the rear, Skorpion ready, but no one followed. Gravel became tar, intersecting a road.

"Southeast," she said. "There'll be a town sooner or later. If we go north, we'll hit river. Once we find out where we are, I can send for help."

"Unless we end up where his men are."

"Might have only been those two."

"Not likely. Eight vans kidnapped us. That was a pretty complicated maneuver. Practiced, I'd say. Even if six of the vans were remote servo-linked, responding to the movements of the manned cars, there's still too much room for error. I'd say he's got a small army."

"It must be fascinating to be in your head."

"What do you mean?"

"You see these... pieces. Bits here, bits there. Huge gaps between them, maybe they're not even an obvious fit. But you organize them, fill in the gaps, or jump them. A mosaic of educated guesses. Are you always right?"

"I wish. I just test hypotheses, a bad habit I picked up from my dad. Sorry."

"Don't be. You'd make a good spy."

"Is that supposed to be a compliment?"

"An observation. So what have you gleaned about our host?"

"Besides that he let us go, you mean?"

"You're serious? He kidnaps us so we can escape? That doesn't make sense."

"It's the only way it makes sense. I think he wanted me to escape the night he led me to the cabin. Only I didn't take the bait."

"Why didn't you?"

"Curiosity. I wanted to know more. He planned to kill you anyway, staying around seemed the best way to keep you alive."

"Thank you. This is insane, you know."

"Expecting me to believe he wanted me to work on something, that was insane. The parts and equipment he got me were kid stuff. About the only thing I could've built with them is the damper box, and that only worked because I pilfered parts from his bugs. It was sham work, Jean."

"I still don't understand."

"Follow the chain: James Creed is murdered. Why? Either someone does not want him revealing something to the Man of Iron, or he had already been eliminated as an information source. Unless it was totally unrelated, but let's dismiss that for now. Whatever Man of Iron wanted, he couldn't get it from Creed. Creed's research partner was my grandfather, and suddenly I'm involved. He goes to a lot of trouble to start me thinking about what my grandfather knew. Ergo: whatever he wants, he thinks Grandad had. And he thinks I can help him."

"So he grabs you, plays out this little charade, and lets you go?"

"I'm a Judas goat, remember. What put your people onto him?"

"We got a tip. Anonymous information. They said he was working with Neo-Nazis."

"I think he tipped you. It's you he's been working with, whether you knew it or not. You did everything he expected. I wouldn't be surprised if he has a mole in your organization."

"That theory will thrill Dad no end."

"Try this on: how did he even know about Creed and Grandad, and me? They were involved in a secret war project. Was it ever declassified? If it was, I never read anything about it."

"No. It wasn't."

"Maybe that's the whole point of this exercise, to smoke out the mole."

A truck rolled up. She threw the gun in bushes, and waved.

⟨⦁⟩

The restaurant featured American fast food dressed in Swiss tradition. Zac sipped a shake, watched Jean through the window, on a phone, animated. She crossed traffic and entered.

"There's a cemetery not far from here. That'll be our pickup point. We're to wait at the grave of a Karl Hagen."

Sirens. Firetrucks rolled past, heading north.

"Any reason?"

"A standard meet spot. It's been in the books

for decades. Apparently, he was one of ours. Killed back in the war. McBane was always sorry about that. Some sort of hero, I'm told, but they won't tell me how."

"That's curious, given the circumstances: us here, with leftover problems from WWII, right where one of your father's agents died at that time."

"There you go. Always thinking."

"It's what gets me into trouble." Slurping the dregs.

⚡⚡⚡

Tombstones, basic and indistinguishable, surrounded them. Chalk-white stripes across the glen. "There must be thosands!"

"Hundreds." Jean, checking he watch.

"The newer arrivals will be closer to the edges, I should think.

We're looking for 1945."

Eight minutes, they found the first rung. Twenty minutes.

"Hagen, did you say? Jean, I found him."

A fly on the stone. He caught and crushed it. Pulp, no microchips. He scanned the darkening forest.

"What are you looking at?"

"This is it? We get picked up and we're gone? It doesn't fit. He's watching."

"He's probably still paralyzed where we left him. No one followed us. No one's watching."

"He is."

Shuddering wind: the helicopter, alighting. A rope ladder hit the ground. Inside: a dark-faced man, half masked by a helmet. Walther in hand. "Code word?" over the roar.

"Golden Boy," Jean shouted.

"Names!"

"Severn. Robillard."

"What was that last one?"

"ROW-BILL-LARD." The earth shuddered, subsided.

"Check." The Walther disappeared. "Come on up."

Jean started up the ladder. "Zac! Hurry up."

"Go ahead."

"Zac!"

"I'm staying. Trust me." Hand out, his fingers splayed in a signal. Five. He repeated. She nodded and climbed, and the helicopter rose from sight, staccato wind fading and distant.

Alone, in the dark. On the tombstone, Karl Hagen, ?-1945. Simple, what else was there to know? Talk to me, Karl. What does he want?

Metal feet impacted grass. Zac closed his eyes, didn't turn. His hand shook. Behind him, the smoker's drone, unmodulated anger: "Robillard!" An upward ripple beneath Zac's feet. "What have you done?"

Catching his breath, Zac turned. "I'm not playing anymore. Tell me what you want or kill me, but I won't be your goat."

"Your grandfather stole it from me. Return it!"

"It?"

Metal fingers choked him, clutched his throat. "He built the robot! He programmed it! It would have returned to him! A man, without patience or time. I need that data, Robillard." Ground, or his knees, shook beneath him.

"The Golden One. Golden Boy?

He was never found?"

"No more chances." Man of Iron lifting him, by the neck, off his feet. In his ear, a whisper from long ago: Robillard.

"Robillard!" he shouted. Man of Iron stopped squeezing, puzzled. Again: "Robillard! Help me!" With the last of his air, the loudest: "ROBILLARD!"

An eruption, beneath them: the coffin, exploded from the ground. Man of Iron lost his footing, his grip slipped. Zac fell. The coffin dropped, split, pocked bones rolled: the last of Karl Hagen.

Rising from the tomb, dark in mud and moonight: a robot of gold.

"At last." An excited rasp, tinged in self-pity. "Here, all this time." Golden Boy moved stiffly, a cartoon. With ease, Man of Iron grabbed a golden arm.

A bullet spanged off him, then another. Overhead, the helicopter, circling back. The Man of Iron ignored them, snapped a golden leg in two.

Over the pounding blades: "Zac!" Jean, holding the Walther.

"Jean! The gun! Give me the gun!"

He caught it, spun. Man of Iron laughed, the golden robot truncated, iron hand over its face. The first shot hit mud, the second an iron shoulder.

An iron growl: "Nuisance." Contempt in the voice, as the monster shrugged off a third shot. Behind Zac, the ladder dangled. He ignored it, trembling. Man of Iron striding. Almost a run. An arm sweep, he stepped back as metal fingers grazed his nose.

He fired. Six rapid shots, locked arms, inches from the iron face. The robot gave a rasped howl. Man of Iron staggered, hands over his eyes.

Zac tossed the gun. He snagged Golden Boy on the rope ladder, and held as the copter rose.

From the forest, a mob came, screaming, shooting: Man of Iron's army. Zac cowered behind the wrecked golden shield. Then they were out of range, Jean at the controls, the pilot helping him up.

"What did you do?" Jean, when they were safe inside.

"Shot him in the eyes."

"You blinded him? That's brilliant."

"I doubt I hurt him at all. But I told you, whatever he is, he's a man. He still behaves like a man. Basic reflex, back off when your eyes are attacked. I only needed to buy a few seconds."

"Like I said, you'd make a good spy."

"Please stop saying that."

The pilot, reticent: "I flew over the house where they held you. It's ashes, now, burned to the ground."

Jean sighed. "He's covering his tracks. We're back to square one, now."

"Not quite." Zac, patting the golden head. "His secrets are somewhere in here. I'm going to find them."

He glanced out the window. Below, black dots reddened in the dim moon.

PART THREE: JEAN
England

Zac Robillard sat drenched in sweat, too aware of his clothes sticking to his skin. Three days he had been back in England, with no rest and barely any food, answering question after question, and still his inquisitor cooly paced without pause or reaction.

"I don't know you," he had said on his arrival, when Severn's agents swooped onto them at Gatwick Airport to leave him in hawknosed Fenniman's custody and whisk Jean and the remnants of Golden Boy away. Fenniman, whose frozen lips never frowned or smiled, had replied:

"I'm not someone one wants to know, sonny Jim. Do as you're told."

A whirl of guards and opaqued glass—look for a car with opaqued glass and you'll find a spy, Zac reasoned—a treacherous winding drive, and a room too brightly lit for sleep, that was everything he had experienced in 72 hours, those and the blurred memory of his own voice, broken by the occasional question or Fenniman's momentary absence. Robillard felt unreal, a walking dream, running and running but never gaining ground.

Fenniman's pacing continued. "Start from the beginning. I want to hear it again."

"I've told you everything."

"Can't be too thorough. Again."

⊶⚬⚭⚬⊷

"Dad, stop this," Jean said. On the monitor, Zac's speech stumbled, his English barely recognizable.

"Fenniman's right, Jean. With what's on the table, we can't be too careful."

"Question me, then."

"Jean, I trust you."

"Is sentiment part of the job now? That's not what the instructors taught us. The security of the realm precludes sentiment, remember?"

"Jean..."

"Interrogate me."

"No."

"Then let him go."

"Not until we're certain he's not in league with the enemy."

"I'm certain. Do you trust me or not?"

"We cannot take the chance."

"If you don't trust me, interrogate me."

She had her mother's skill at painting him into a corner. Reluctantly, he tapped his intercom.

"Fenniman, that's enough. Get him cleaned up."

⊷ ⊶

The shower smelled of must, caulk edges mildewed black by generations of English dampness. The building, now that he saw it, reeked of age: an old manor of 18th-century design, kept up with conveniences like indoor plumbing added but essentially unchanged in three centuries. Zac dried himself and toweled off his hair, then slipped into the thick flannel robe left for him.

Jean was waiting for him when he entered his room.

"Where's Fenniman?" he asked.

"On break. Are you all right?"

"I was hoping to get some sleep. What do you want?" She looked hurt, but he was tired. Civility was beyond him for the moment.

"To apologize. To thank you." She stood and kissed him, lingering before backing off. He watched her in disbelief. "What's wrong?"

"I never know what game you're playing," he said.

She turned her face away, almost shy. "No, I don't suppose you ever will."

"When can I leave?"

"That's up to my father. You'll be meeting with him soon."

Then she was gone, slipping from the room before he could think. Slowly, he licked the taste of her from his lips, and decided to follow, but the door had locked shut behind her.

He fell on the bed to sleep, but the image of her face danced inside his eyelids, keeping him awake. With great effort and buoyed by exhaustion, he clawed his way down into darkness, but her face stayed with him in dreams.

Fenniman ushered him into the office and pointed to an old mahogany captain's chair, next to a matching desk. Zac sat there, hands in his lap, as certain as he had been at the Man of Iron's house that he was being watched. Fenniman left. Minutes crawled by. The desk was empty of clutter; a pen and penknife sat at the head of a green desk blotter, a plain desk calendar was opened to the current date. There were letters in a desktop inbox, face down, and nothing in the outbox. Unlike his own room, there was no noticeable dust, and nothing but a modern telephone to suggest the office had been occupied anytime in the last half-century.

On one wall hung framed photographs, the only personal touch Zac could spot. The frames were as plain as the rest of the room, narrow chrome strips, and the photos small and faded. He couldn't make them out. With a shrug he stood, deciding to ignore Fenniman's order. What could they do to him? Lock him away? He realized whatever power they had over him was what he surrendered, and he was no longer in the mood for it.

The photos were puzzling: Jean with a child in her arms. Jean and himself and a broad-faced stranger clustered in a stiff pose. Jean, older, in a wedding dress, cutting a four-tiered cake with a tuxedoed man. Jean was too old, older than now, all the shots anachronous. The clothes were wrong, the photographs black and white, which no one used anymore: the 40s. On closer inspection, he identified "himself" as his grandfather, and began to find in the broadfaced man a hint of the features of James Creed. Which made "Jean" her mother, Kathleen, and the child in her arms Jean.

It was like peering into a parallel world.

"A handsome clan, eh?" Severn stood in the doorway. His face had thinned and his stomach widened, and his hair was gone, but Zac made him for the groom in the wedding photo. He was shorter than Zac expected, shorter than Zac, and dressed in a nondescript gray suit. Had Zac passed him on the street, he wouldn't have seen him.

Which, Zac suspected, was what one looks for in a spy.

Severn tapped the photo of mother and child. "My favorite," he said, unexpectedly bittersweet, and took a seat behind his desk. "Jean was a wonderful child. Would you care to sit?"

"I'd rather stand."

"You'd be more comfortable sitting."

"I'd rather stand."

"Please," Severn said impatiently, a ghost of a threat in the word.

Zac sat.

"You're very much like Ben. It was always entertaining, watching him clash wills with Ian McBane."

"Who won?"

"It was wartime and Ian had urgency on his side. No matter how good, your grandfather's arguments always crashed on the threat of Nazi victory."

"And now?"

"Funny you should mention it. I wanted to discuss the Iron Major."

"The Iron Major?"

"I gather he retooled himself for you. 'The Man of Iron' is he now? He must have a blacksmith somewhere we don't know about."

"I already told Fenniman everything."

"Yes, I know. I was listening. This isn't another interrogation, Zac.

May I call you Zac? Think of it as a debriefing. No pressure. When we're done here you can be on your way."

"I'd like to go now."

"When we're done."

"Now. In the last ten days, I've been suspended from school, shot at, kidnapped, bounced all over Europe, deceived right and left, virtually enslaved, attacked, dangled from a helicopter, and most recently starved and driven to exhaustion. I'm sick of it. I'd like to get a hamburger and watch a movie. The last thing I want to do is have another idiotic 'chat' with another secret agent. I want to get back to the real world."

"Sorry you feel that way. You won't last out there."

"What do you mean?"

"He's bound to have people looking for you. Without us, you'll be back in his company before you can blink twice."

"You didn't get him."

Severn scratched the corner of his mouth with his thumb and eyed Zac for a long moment.

"Let me tell you a story," he said.

As the Second World War wound down, rumors persisted of desperate German measures to gain victory. My mentor, an American of Irish descent named Ian McBane, collected such rumors, and spread his web of agents across Europe to ascertain the truth of them. These were the unsung heroes of the war, men who fought battles that will never be officially acknowledged. Some worked directly for the German chiefs of staff and passed us important information. Two hindered development of nuclear weapons at Pennemunde. Another swayed some powerful Nazis away from traditional science, encouraging them instead to waste their rapidly dwindling funds on such nonsense as an expedition to the North Pole to find a hole to the center of the earth and the great mythical civilizations supposed to exist there.

McBane's obsession with German secrets began before the war. He never told me why. I know he had already enlisted a small and unusual group of agents, men accepted as world travellers—explorers, doctors, geologists—to travel through Europe and absorb what information they could. Among these was an actor who vanished in Nazi Germany in the mid-30s. Through later reports, McBane pieced together the events: the actor had been arrested without cause on a train and surrendered to a state-funded scientific project conducted in an old castle in Eastern Bavaria. Whatever happened to the actor, the project was abruptly halted when the castle was destroyed. McBane had his suspicions of what happened, but nothing was ever proved. The brief chaos following the destruction cracked the previously impenetrable security surrounding the project, however, and the information McBane sought quietly leaked out. The report struck terror into him.

The Nazis, under two geniuses named Grosswald and Muller, were developing a mechanical man.

The American government dismissed McBane's report. At that time, Hitler was still viewed by many as a strong but reasonable man, and if a few people were dying because of him and his followers, well, it was a German internal matter, and, in any case, they were not people who mattered to anyone. Undaunted, McBane approached England, which had more immediate fears of a powerful Germany. Prime Minister Chamberlain was also disinclined to interfere in matters of the German military, as he, along with many other Englishmen and Americans, viewed a strong Germany as Europe's bulwark against Soviet Russia and secretly believed that the two powers would destroy each other, leaving the West unscathed.

But a series of social encounters resulted in McBane's introduction to King George, and after listening to McBane's stories, the King took it upon himself to organize a similiar project, ostensibly purely scientific in purpose. Through the auspices of the Crown, pressuring your President Roosevelt to equally involve American resources, a Western plan to develop a mechanical man was put into motion.

It was now 1938. McBane had spent over two years attempting to awaken the sleeping Western powers, and now Hitler was muttering covetously about reclaiming a mythical motherland in the east of Europe, and the wind stunk of war. England and America scoured their finest scientists and scholars, and each recruited one expert, if one can be called expert in a field that doesn't exist. From the British Royal Academy Of Science came James Creed, whose skill with complex mechanicals was unparalleled, and from America, the brilliant Benjamin Robillard, a former neurosurgeon fixated on the possibility of artificial replications of the human brain. Their initial research took place at a small college in Massachusetts, where once a month they masqueraded as visiting lecturers.

The initial work did not go well. Creed was dismissive of his new surroundings, feeling America to be a cultureless and savage place, while Robillard's attitude was far from good. Not long before, in Illinois, his wife had died of pneumonia, and as capable a scholar as he was, in his melancholy he found himself incapable of caring for his young son alone. He was depressed and despondent, and when the offer came from the government, he saw it as a means to gain a new purpose. Reluctantly, he left the boy with his aunt and went East, but for more than a year, though he and Creed formed a fast friendship, his depression did not lift, and little was accomplished. It was thought, to McBane's chagrin, that the project was a failure and would need to be shut down.

It took a visit with his son to reverse Robillard's attitude. The boy met his father in New York City, where they took in the World's Fair. Its builders had thought to peddle a new future to the world, but after almost a decade of depression, with war rumbling around the corner, what they were really peddling was hope, and hope was what Robillard desperately wanted to buy. Here was a gleaming future of marvels and vigor, which suggested that the future not only existed but that it would be better than the present, ineffably better: nothing less than the total victory of human achievement. Had he been alone, he might not have been so inspired, but he witnessed the awe in his son, and was infected by the boy's dreams. He had been wracked with guilt at abandoning the boy, but here was the chance to make that sacrifice mean something, to help to build the future the boy clearly craved. There at the fair they found a symbol of that future, a mechanical marvel that shined like pure gold, and Robillard only half-mockingly referred to it as Golden Boy.

Their weekend was over, and the son returned to the Midwest while Robillard went back to Massachusetts. He and the boy were reconciled, and he had rediscovered his vision. His life had purpose again. But it was too late.

Two weeks after Robillard's trip to the fair, Hitler invaded Poland, and England, responding to old pacts, launched into war. Russia was at truce with Germany, Italy allied with it, Spain in sympathy, and the Nazis swept like lightning over Continental Europe, conquering France and the Low Countries almost overnight. America remained neutral. In all the west, it seemed as if only tiny England stood against them.

And the project was over. Britain needed Creed back for other things, and there were other places to spend English money. Creed, being an English patriot and sick of Massachusetts, leaped at the chance to return home. In America, economic recovery was slow, and secret projects were difficult to justify. Before Robillard could accomplish anything, his research was confiscated and boxed away, and he was sent packing.

McBane received an invitation from the Crown to lend his espionage skills to the English cause, and, finding himself increasingly in disfavor in American political circles that mostly wished to remain neutral in the war, he accepted.

Things changed again in 1942, after America was drawn into the war. Suddenly McBane was in charge of an expanded spy network, with fresh recruits continuously drawn from American sources, and soon he had enough agents to return to his original obsession, the tracking of German scientific efforts. Still feeling guilty over the death of the actor five years before, McBane assigned a special agent simply to watch in the vicinity of Castle Grosswald for any continuation of the robotics program. There were suggestions it had been revived at the height of Hitler's power, and by 1943, McBane had persuaded his English and American masters to renew their own project, once again bringing together Robillard and Creed, along with Creed's young Scottish assistant, the lovely Kathleen Reay. Lacking a laboratory, he set them up in his own headquarters, at Hembley Manor, in the English countryside, the summer residence of a cousin of the King. Again they worked with scant resources, but both scientists had continued their research on their own, and felt on the verge of a breakthrough.

Then reports began filtering in that the Germans had surreptitiously moved a large equipment caravan from Eastern Bavaria to southwestern Wurttemberg, and over the Swiss border. If they were attempting secrecy, the parade was comical: advanced machinery sloppily covered with tarps and carried cross-country on the backs of horse-drawn carts. But there was little else left in Germany they could use. By chance, McBane had an agent placed in their vicinity in Switzerland, a young American named Elliot Hecht, who worked under the name of Karl Hagen. Hecht confirmed McBane's fear that the Nazi robotics program was well underway, and described an encounter with a fearsome, apparently self-directed monster built entirely of iron.

To Robillard, the story was nonsense. He and Creed had tried sporadically for almost eight years to accomplish the simplest robot without success and with the most modern technology available. Robillard, whose field was neurology, had finally conceived an approximation of a human brain that might store sufficient levels of information to allow a robot to function at all, and he could not see that a German working with far inferior technology could possibly have accomplished more.

But McBane told them he needed a robot and he needed it then.

So, Robillard had a brainstorm.

They needed a mechanical body to house their experimental artificial brain, and he recalled the Golden Boy from the fair. It seemed he was the only one of his kind. McBane's agents found the mechanical man abandoned in a warehouse full of fair refuse. The fair had closed after 1940, and with the war, that imagined perfect future had been forgotten.

The secret of Robillard's 'new brain' was known only to him and Creed, and their prototype was fitted into the Golden Boy's metal skull, and attached to his mechanicals with such improvements as Creed could quickly install. It was determined

Golden Boy could store and process a finite set of commands, and so a key command was also installed, a single word that would trigger either action or a memory dump in preparation for new commands. Virtually untested, the creation was flown to Switzerland, where it and The Word were delivered to Hecht. The Golden Boy and the Iron Major met, that much we know, and the German project was destroyed. Hecht was found dead shortly afterward, and buried in the nearest village, according to last wishes found in his room. The whereabouts of either robot was unknown. They were both presumed to be destroyed with the project, but no evidence existed to support that theory.

Nazism did not die with Nazi Germany. It existed in silence for many years in many places throughout the world. It took a generation who had no direct experience of the war to resurrect it to any degree, influenced directly or from afar by those who had fought the war and lost. It was rooted in small pockets, in Germany, yes, but in England and in America as well. Nazi factions stripped themselves of their ideological pasts and made themselves useful to various governments, and in that way spread their disease. But the groups were spotty, as likely to feud with each other as to grow into general threats, and in this way they rendered themselves impotent.

Yet only a few years ago, these rivalries and divisions seemed to fade away. The factions were melding into an underground force. Something had galvanized them, given them a center to focus on. Our first suspicion was a high-ranking survivor of the war, making himself a rallying point. Our intelligence revealed all known survivors accounted for and out of commission. While holding a symbolic value, they were too old for the young to take seriously. No one cared for defeated tales of the old days. Something was setting their sights on tomorrow.

Slowly, rumors began to filter out. Known neo-Nazi leaders made pilgrimages, meeting with a mystery man, and the stories were that the man was made of solid iron.

Finally a photograph was snapped, a fleeting glimpse of the Man of Iron. It seemed impossible. Excavation crews were secretly sent to the scene of his alleged demise, a cave, where they found the remains of the original robot project. And signs that something had climbed from a deep crevice, digging handholes into solid rock with its fingers. A tiny tunnel was smashed through the side of a mountain, and experts guessed it had been created over decades, though they could not guess the method. No sign of Golden Boy was found. A picture began to develop: the Iron Major cast into the crevice, spending years relentlessly clawing and scraping his way to freedom. There was no sign of anything having lived in the cave in all that time, with the conclusion that the Iron Major was indeed acting on his own impulses. Finding the Reich destroyed, he set about to unite disparate bands into a single force, with himself as symbol of the indestructibility of that twisted ideal.

Even with what little we learned about him, it became clear that he was not yet ready to move. He was waiting for something, even quietly hunting for it. No matter how hard we tried, we could not find out what that was. While we watched, his forces strengthened and grew, waiting to be triggered, and for the new world of horror to begin.

Zac sat silent, weighing Severn's words. At last, he said, "Everyone has a story for me."

"You don't believe me?"

"No, you're probably telling the truth. I want to see Golden Boy."

"Not possible. Security."

"Make an exception."

Severn shook his head.

"The Man of Iron wanted Golden Boy."

"We know. Jean told us."

"But you don't know why he wanted him."

"You do?"

"No. I'll ask Golden Boy."

"And how do you propose to do that?"

"I know the magic word."

⬧

Fenniman took him to the special room. Concrete walls six-feet thick, no windows, timelocked steel door; Zac felt as if he were back in the Man of Iron's house. It was a prison. Sodium lights blazed overhead, shining off the tile floors, not a shadow in the 6x6 foot room. Tools hung from a pegboard mounted on the back wall. Two television cameras angled from the corners down to the steel table in the center of the floor, as in an operating theater. He glanced at the TV screen embedded in the wall, the only interface with the outside once the door closed. Over the door was the digital display of a three hour timer, counting down the time lock.

On the table lay the truncated remains of Golden Boy.

"No one's touched it?" Zac asked.

"They're afraid to. Not keen on damaging it. Me, I'd just take it somewhere and blow it up."

"No doubt. I need to do this alone."

"Not a bloody chance, son. I don't put sabotage beyond you."

"Then we learn nothing."